T0137644

Hyperspectral Remote Sensing
Principles and Applications

Taylor & Francis Series in Remote Sensing Applications

Series Editor

Qihao Weng

Indiana State University
Terre Haute, Indiana, U.S.A.

Hyperspectral Remote Sensing
Principles and Applications

Marcus Borengasser
William S. Hungate
Russell Watkins

CRC Press
Taylor & Francis Group
Boca Raton   London   New York

CRC Press is an imprint of the
Taylor & Francis Group, an **informa** business

Cover imagery used with permission from ITRES Research Limited, Alberta, Canada.

CRC Press
Taylor & Francis Group
6000 Broken Sound Parkway NW, Suite 300
Boca Raton, FL 33487-2742

© 2008 by Taylor & Francis Group, LLC
CRC Press is an imprint of Taylor & Francis Group, an Informa business

No claim to original U.S. Government works

ISBN-13: 978-1-13874-718-0 (Paperback)

Library of Congress Cataloging-in-Publication Data

Borengasser, Marcus.
    Hyperspectral remote sensing : principles and applications / authors, Marcus Borengasser, William S. Hungate.
        p. cm.
    Includes bibliographical references and index.
    ISBN 978-1-56670-654-4 (hardback : alk. paper) 1. Remote sensing. 2. Multispectral photography. 3. Image processing. I. Hungate, William S. II. Title.

G70.4.B68 2007
621.36'78--dc22   2007030243

Visit the Taylor & Francis Web site at
http://www.taylorandfrancis.com

and the CRC Press Web site at
http://www.crcpress.com

Contents

Contents

Preface

Many people today are in management positions that require an understanding of complex technical issues before the best decisions can be made for their company or organization. This hyperspectral remote sensing book is written at a level that can be understood by those who deal with land management issues such as mapping tree species, identifying invasive plants, and identifying key geologic features.

The first half of the book explains the basic concepts and underlying principles that lead to the creation of a remote sensing image. This section will take you through all the major aspects of hyperspectral image acquisition, exploitation, interpretation, and applications. First, the introduction to spectral radiometry presents concepts such as radiance, irradiance, and flux; blackbody radiation; and atmospheric interactions. Next is a discussion of imaging spectrometers, including an explanation of spectral range, FWHM (full width half maximum), resolution, sampling, SNR (signal-to-noise ratio), and multispectral and hyperspectral sensor systems.

Following this introduction of how a remote sensing image is constructed is a series of chapters on information extraction. You will learn the underlying physics principles that lead to the creation of the image and how to interpret the information in the images.

The second half of the book describes case studies that have applied this information to the use of hyperspectral remote sensing in agriculture, forestry, environmental monitoring, and geology. The case studies in each chapter illustrate how hyperspectral remote sensing is being used to solve many of the land management issues that confront our society.

Topics for agriculture, forestry, and environmental monitoring applications include detection of crop disease, crop growth analysis, classifying water quality, mapping submerged aquatic vegetation, and estimating hardwood chlorophyll content. For geology applications, topics include detecting hydrocarbons and identifying and mapping hydrothermal alteration.

This book is designed to be used by people who have not used hyperspectral data but realize that hyperspectral technology may offer a solution to their application area. After reading this book, you will have a better understanding of how to evaluate different approaches to hyperspectral analyses, as well as which approaches may or may not work for the applications of interest to you.

1

History and Description of Hyperspectral Imaging

Origins

And then there was light. And since the beginning of time, light was thought to be pure and singular. It wasn't until our ancestors started exploring celestial and terrestrial objects that scientists started to understand light and its properties.

No one knows when the first glass prism was made, but in the beginning, people came up with many interesting but incorrect concepts to explain its effect on light. There was no question that when light passed through a prism it produced a spectrum of colors, as shown in Figure 1.1. People thought that the impurities of the glass caused this display of colors seen in a rainbow. But in 1666, Isaac Newton proved the glass was not the source of the colors, but the natural light itself.

His experiment was simple. He used one prism to separate the light into the color spectrum. He then used a second prism on each of the individual colors. When he showed that the light coming out of the second prism was the same as the light that went into that prism, he had proven that the prism was not the source of the colors.

So then scientists knew that natural light contains colors. But what is light? What causes the colors? Is light a particle or wave? These questions led to some of the greatest discoveries made by Newton and other researchers in modern physics and electro-optics.

Much of Newton's greatest work was published in *Principia Mathematica* in 1687, but he still considered light to be a bullet-like particle that traveled in a straight beam, in contrast to scientists like Christian Huygens who had theorized that light moved in waves. It was not until 1803 that Thomas Young was able to prove the motion of light followed wave patterns (Young, 1804).

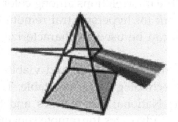

Figure 1.1   Light passing through a prism.

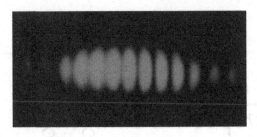

Figure 1.2   Interference pattern.

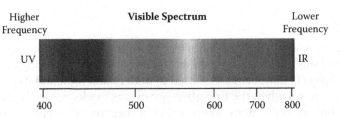

Figure 1.3   Visible spectrum with wavelengths in nanometers.

His experiment is repeated on almost a daily basis in physics classrooms around the world. Called the "double slit" experiment, its simple and clear methodology proves that light has wave properties.

His experiment is based on the theory that interacting waves have constructive and destructive interference, much like when you throw two pebbles in a pond and the waves meet. A collimated beam of light is projected through a tiny hole or shutter. This column of light is then passed through a narrow, vertical slit and projected on a screen or another surface. A representation of interference patterns is shown in Figure 1.2.

At about the same time in the 1800s, William Herschel was also expanding on Newton's glass prism discovery. Herschel observed that the different colors actually varied in temperature. Blues and greens were cooler, while reds were warmer. In fact, he further discovered that the energy present outside the red spectrum was warmer yet, and this region became known as infrared. This was the first documented reference that the visible light spectrum was related to frequency and wavelength, as shown in Figure 1.3.

These discoveries of the interrelations among color, frequency, and wavelength laid the framework for hyperspectral remote sensing because these fundamental principles can be used to characterize the reflection of light against objects.

Once modern aviation became a safe and viable platform, the integration of remote sensing technology was inevitable. The U.S. Department of Defense (DoD) and the National Aeronautics and Space Administration (NASA) have been strong advocates for remote sensing and have sponsored the development of many systems.

From the beginning, airborne imagery of terrestrial objects was a vital form of information to detect objects and features on the ground. Monochrome and panchromatic cameras were the first class of remote sensors used for aerial photography, reconnaissance, and surveillance in both civil and commercial capacities.

In the 1960s, the DoD retrofitted the infrared film cameras on the U-2 high-altitude reconnaissance plane with special spectral filters. These cameras were among the first multispectral cameras in existence.

In July 1972, NASA launched the Earth Resources Technology Satellite (ERTS), which later became Landsat 1. The system was designed for extensive mapping and remote sensing of the Earth's surface and was the first system capable of producing multispectral data in digital format. The applications of Landsat imagery have been demonstrated in agriculture, cartography, environmental monitoring, forestry, land use planning, and oceanography.

With the innovation of faster computers that could handle the enormous amount of data required from new and improved spectrometers, hyperspectral remote sensing has flourished in the defense and commercial sectors.

Definitions

The term "multi" is derived from the Latin word for "many" and "hyper" is the Greek word for "over," "above," or an "exaggerated amount." These, combined with "spectral," which relates to colors, are combined to form "multispectral" and "hyperspectral," which figuratively mean "many colors." The science of multispectral and hyperspectral remote sensing is based on taking a portion of the electromagnetic spectrum and breaking it into pieces for the purpose of analytical computations.

The following basic terminology is defined to further explain the basics of the electromagnetic spectrum and hyperspectral remote sensing.

Photon

A photon is a discrete particle of electromagnetic energy having no mass, no electric charge, and an indefinite life. The existence of photons was first based on the interpretation of experimental results and presented in a scientific paper by Albert Einstein in 1905.

The energy E of any photon is related to its frequency as follows, where h is Planck's constant (6.626068×10^{-34} m^2 kg/s) and υ is the frequency:

$$E = h\upsilon$$

Electromagnetic Spectrum

Each photon of the electromagnetic spectrum has a wavelength determined by its energy level. Light and other forms of electromagnetic radiation commonly are described in terms of their wavelengths. The visible spectrum

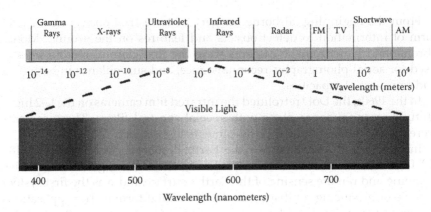

Figure 1.4 Electromagnetic spectrum.

is shown in Figure 1.3. Its relationship to the entire electromagnetic spectrum is detailed in Figure 1.4.

Speed of Light

Up until the 17th century, scientists believed that light was instantaneous and could travel any distance in no time. Galileo disagreed and tried to disprove the theory with shuttered lanterns on adjacent hilltops. He failed—not in theory, but in execution. He was too close to quantify the measurement he was seeking.

In 1675, Danish astronomer Olaf Roemer was making precise measurements of the orbit of Io, one of Jupiter's moons. With these observations, Roemer should have been able to predict Io's location precisely. But over the course of time, when he realized the measurements seemed inaccurate, he was able to relate the inaccuracies with the time of year. Io seemed to be ahead of schedule when Earth was closer to Jupiter and behind schedule when Jupiter was farther away. He began to wonder if the reflected light of Jupiter and Io took time to travel. Roemer then became the first to calculate the speed of light at about 186,000 miles per second. Since then, more precise instruments and methods allow us to measure the speed of light, like placing mirrors on the moon and using the laser time-of-flight to further refine the speed, but Roemer wasn't that far off.

The designation for the speed of light is c and is related to the frequency υ and the wavelength λ in any part of the electromagnetic spectrum.

$$c = \upsilon\lambda$$

Emission and Reflection

Photons can be absorbed, reflected, or transmitted. In the realm of thermodynamics, radiated heat creates photons. Radiation heat transfer is the

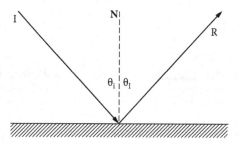

Figure 1.5 Specular reflection.

exchange of thermal radiation energy between two or more bodies. Thermal radiation is defined as electromagnetic radiation in the wavelength range of 0.1 to 100 microns.

Radiation heat transfer must account for both incoming and outgoing thermal radiation and can be expressed as:

$$1 = \varepsilon_{reflected} + \varepsilon_{absorbed} \; \varepsilon_{transmitted}$$

Because most solid bodies are opaque to thermal radiation, transmission can be ignored.

$$1 = \varepsilon_{reflected} + \varepsilon_{absorbed}$$

To account for emissive radiation, a comparison is made to a perfect blackbody, which is a theoretical object that absorbs 100% of the incident radiation, reflects none, and appears perfectly black. The ratio of the actual emissive radiation E to the emissive power of a blackbody is defined as the surface emissivity ε.

$$\varepsilon = E/E_{blackbody}$$

Reflectance is the percentage of incident light that is reflected by a material. In climatology and remote sensing, reflectivity is commonly referred to as "albedo," the Latin term for white.

Reflectivity R can be expressed as:

$$R = 1 - \varepsilon$$

Surface reflections can be specular or diffuse. Figure 1.5 depicts specular, or mirrorlike, reflections where the incident wave I of energy against the surface is the same as the reflected wave R. The incident and reflected angles are denoted as θ. N is the angle normal to the reflecting surface.

The difference between specular and diffuse (or Lambertian) scattering as shown in Figure 1.6 is based on surface roughness and granularity.

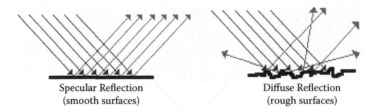

Specular Reflection   Diffuse Reflection
(smooth surfaces)   (rough surfaces)

Figure 1.6   Specular vs. diffuse reflections.

As surface roughness increases, so does the amount of diffuse scattering. The size of the particles on the surface, not the quantity, contributes more to the diffuse scattering. A desert or beach with small grains of sand may act almost as a specular surface.

Emissivity and reflectivity are two of the fundamental physics principles that govern hyperpsectral remote sensing. The portion of the electromagnetic spectrum sought for exploitation from a hyperspectral sensor—visible, near infrared, shortwave infrared, midwave infrared, or longwave infrared—will define the design specifications for sensor optics and detector array, as well as the application of the remote sensor.

Reference

Young, T. 1804. The Bakerian lecture: Experiments and calculations relative to physical optics, *Philosophical Transactions of the Royal Society of London*, 94, 1–16.

2

Spectral Radiometry

Any hyperspectral data collected from an airborne or spaceborne platform is influenced by the atmospheric conditions at the time the data is collected. If the weather is sunny and clear, the data will be optimum. Data collected when the atmosphere is cloudy or humid, however, will be affected by those conditions. Therefore, knowing the principles of spectral radiometry and understanding how to use its concepts is important when interpreting the data collected by hyperspectral sensors.

Principles of Spectral Radiometry

Radiometry is the physical measurement of electromagnetic radiation within the ultraviolet, visible, and infrared wavelengths. A radiometer is a device used to measure the radiant flux or power in electromagnetic radiation. The most important characteristics of a radiometer are spectral range (wavelengths measured), spectral sensitivity (sensitivity vs. wavelengths measured), field of view (18 degrees or limited to a certain narrow field), and directional response (typically the cosine response of the unidirectional response).

Radiometers can use all kinds of detectors. For example, thermal detectors absorb energy and convert it to a signal. Photon (photodiode) detectors have a constant response per quantum (light particle). The radiation detector within a radiometer is usually a bolometer, which absorbs the radiation falling on it and, as a result, rises in temperature. This rise can then be measured by a thermometer. This temperature rise can be related to the power in the incident radiation.

Solid Angles

The radiant intensity describes the flux per unit solid angle from a point source into a particular direction. Although the intensity provides directional information, it does not provide any spatial information. The simplest term used to describe directional or dispersive information involves the solid angle.

A plane angle is the angle formed by two lines meeting in the same plane. Plane angles are measured in either degrees or radians (Figure 2.1). The abbreviation for the radian is rad. Because a circle has 2π radians, the conversion between degrees and radians is 1 rad = $(180/\pi)$ degrees.

A solid angle (Figure 2.2) extends the concept of a plane angle to three dimensions. The solid angle is the ratio of the spherical area to the square of

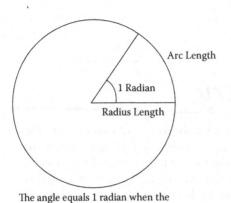

The angle equals 1 radian when the
arc length equals the radius length

Figure 2.1 Definition of a radian.

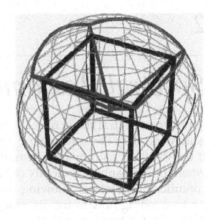

Figure 2.2 Solid angle cube
(Weisstein, 2005).

the radius. The spherical area is a projection of the object onto a unit sphere, and the solid angle is the surface area of that projection. When you divide the surface area of a sphere by the square of its radius, the result is 4π steradians of solid angle in a sphere. One hemisphere has 2π steradians.

Radiance, Irradiance, and Flux

Radiance and spectral radiance are radiometric measures that describe the amount of light that passes through or is emitted from a particular area and falls within a given solid angle in a specified direction. These measures are used to characterize both emission from diffuse sources and reflection from diffuse surfaces. The SI unit of radiance is watts per steradian per square meter ($W \cdot sr^{-1} \cdot m^{-2}$).

Radiance characterizes total emission or reflection, while spectral radiance characterizes the light at a single wavelength or frequency. The radiance is equal to the sum (or integral) of all the spectral radiances from a surface. The SI units for spectral radiance are $W \cdot sr^{-1} \cdot m^{-3}$ when measured per unit wavelength, and $W \cdot sr^{-1} \cdot m^{-2} \cdot Hz^{-1}$ when measured per unit frequency interval.

Radiance is useful because it indicates how much of the power emitted by an emitting or reflecting surface will be received by an optical system looking at the surface from some angle of view. In this case, the solid angle of interest is the solid angle subtended by the aperture of the optical system. Because the eye is an optical system, radiance and luminance are good indicators of how bright an object will appear. For this reason, radiance and luminance are both sometimes called "brightness." Although this usage is discouraged, the nonstandard usage of brightness for radiance persists in some fields, notably laser physics.

The radiance divided by the index of refraction squared is invariant in geometric optics. This means that for an ideal optical system in air, the radiance at the output is the same as the input radiance. This is sometimes called conservation of radiance. For real, passive, optical systems, the output radiance is *at most* equal to the input, unless the index of refraction changes. For example, if you form a demagnified image with a lens, the optical power is concentrated into a smaller area so the irradiance is higher at the image. The light at the image plane, however, fills a larger solid angle so the radiance comes out to be the same, assuming there is no loss at the lens.

"Irradiance," "radiant emittance," and "radiant exitance" are radiometry terms for the power of electromagnetic radiation at a surface per unit area. The term "irradiance" is used when the electromagnetic radiation is incident on the surface. The other two terms are used interchangeably for radiation emerging from a surface. The SI units for all of these quantities are watts per square meter (W/m^2). These quantities are sometimes called "intensity," but this usage leads to confusion with "radiant intensity," which has different units.

All of these quantities characterize the total amount of radiation present, at all frequencies. Each frequency is also commonly considered in the spectrum separately. When this is done for a radiation incident on a surface, it is called "spectral irradiance" and has SI units W/m^3, or commonly $W \cdot m^{-2} \cdot nm^{-1}$.

If a point source radiates light uniformly in all directions and there is no absorption, then the irradiance drops off in proportion to the distance from the object squared because the total power is constant and spread over an area that increases with the square of the distance from the source.

Radiant flux or radiant power is the measure of the total power of electromagnetic radiation (including visible light). The power can be the total emitted from a source or the total landing on a particular surface.

Radiance vs. Reflectance

Radiance is the variable directly measured by remote sensing instruments. Radiance is the amount of light the instrument detects from the object being observed. When looking through an atmosphere, some light scattered by the atmosphere will be seen by the instrument and included in the observed radiance of the target. An atmosphere will also absorb light, which will decrease the observed radiance.

Reflectance is the ratio of the amount of light leaving a target to the amount of light striking the target. If all of the light leaving the target is intercepted for the measurement of reflectance, the result is called "hemispherical reflectance." Hemispherical reflectance is a property of the material being observed. Radiance, on the other hand, depends on the illumination (both its intensity and direction), the orientation and position of the target, and the path of the light through the atmosphere.

When the atmospheric effects and solar illumination are compensated for in digital remote sensing data, the result is apparent reflectance. The difference between apparent reflectance and true reflectance is that in apparent

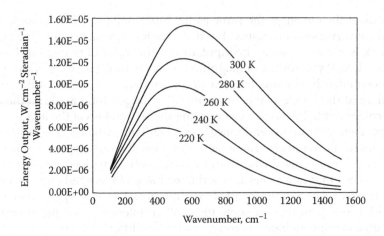

Figure 2.3   Blackbody radiation.

reflectance, the shadows and directional effects on reflectance have not been dealt with (Ray, 1994).

Blackbody Radiation

A blackbody is a theoretical object that absorbs 100% of the electromagnetic radiation that hits it. No radiation passes through it and none is reflected. Therefore, it appears perfectly black. These properties make blackbodies ideal sources of purely thermal radiation because the amount and wavelength (color) of electromagnetic radiation they emit are directly related to their temperature (Figure 2.3).

At a particular temperature, the blackbody emits the maximum amount of energy possible for that temperature. The value of this light is called "blackbody radiation." Blackbodies above 700 K (430°C) produce radiation at visible wavelengths starting at red and ending up at blue as the temperature increases.

Although some materials come very close to being perfect emitters in some wavelength ranges, no real material is a perfect blackbody. Fortunately, during the late nineteenth century, physicists discovered that a cavity with walls thick enough to prevent any radiation from passing directly through them behaved like a blackbody and emitted the same radiation. For this reason, blackbody radiation is sometimes called "cavity radiation." These physicists determined the empirical relationship between blackbody radiation and the two variables on which it depends, temperature and wavelength.

Solar Irradiance and Atmospheric Path Radiance

"Total solar irradiance" is defined as the radiant energy of the sun emitted over the entire electromagnetic spectrum that falls each second on 1 square

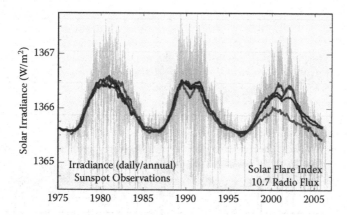

Figure 2.4   Total solar irradiance over a 30-year cycle (Wikipedia).

meter outside the atmosphere of the Earth. In simpler terms, solar irradiance is the output of energy from the sun through the generation and emission of light. Variations in the solar radiance are recorded in the visible, infrared (IR), ultraviolet (UV), extreme ultraviolet (EUV), and x-ray portions of the electromagnetic spectrum. Irradiance is a measurement of the total brightness at any given wavelength. Solar irradiance is the primary source of energy reaching the Earth. Solar irradiation is cyclic, and its correlation to sunspot activity is clearly seen in Figure 2.4.

The measurement of solar irradiance has been conducted from satellites and correlated to ground-based observation of the sunspot activity that generally occurs over an 11-year cycle. Satellite-based observations are depicted in Figure 2.5.

Over the solar irradiance cycle, significant changes occur that impact the type and amount of solar energy that is released. The impact of this change is wavelength and time dependent. Some changes are slight over the cycle. Other changes happen in an order of minutes and can be drastic.

Solar irradiance is directly proportional to sunspot activity. The bright areas around the sunspots are called "faculae." Sunspots, which appear as dark spots on the surface of the sun, are areas of intense magnetic activity that work to block the solar plasma. Sunspots tend to be cooler than other areas, including the faculae. The faculae are responsible for the increased solar irradiance. Faculae are also the result of solar magnetic activity. Figure 2.6 shows an image of sun spots.

The faculae tend to increase and decrease according to the sunspot activity. Even though the increase in solar irradiance is attributed to the increase in faculae when the sunspots also increase, the amount of solar irradiance that reaches the Earth actually decreases during the maximum sunspot activity. On average, the effects of the faculae have more impact on the solar irradiance than the sunspots. Even though the total solar energy reaching the Earth decreases when the surface of the sun that faces the Earth has

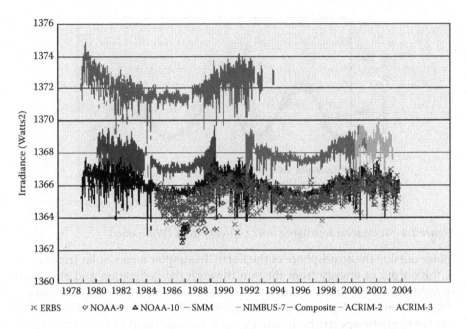

Figure 2.5 Total solar irradiance from the Solar Maximum Mission (SMM) space-craft, from the Earth Radiation Budget Satellite (ERBS), from the NOAA-9 and 10 platforms, and from the Upper Atmospheric Research Satellite (UARS). (Courtesy of National Geophysical Data Center.)

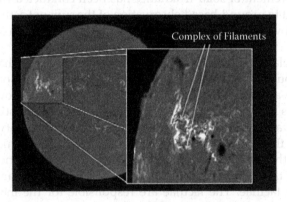

Figure 2.6 Image of sunspots, faculae, and filaments. (Courtesy of Big Bear Solar Observatory.)

many sunspots and faculae, the total energy averaged over a full 30-day solar rotation actually increases.

Atmospheric path radiance is a result of the interaction of radiation with scattering particles in the sky and specularly reflected light from surfaces such as water. Atmospheric path radiance can have an impact to the spectral signature of objects on the ground. The atmospheric path radiance from light scattered from vegetation such as tree leaves can cause some of the spectral properties to be transferred to the next reflective object. This interaction is commonly referred to as "green shine."

Theory of Atmospheric Correction

Atmospheric correction is required to accommodate for the solar radiation interactions with the atmosphere. The quality of the atmospheric correction algorithms directly impacts the quality of the remote hyperspectral data processing and exploitation, and affects the ability to make accurate reflectance measurements from satellite and airborne remote sensors.

Factors that contribute to the successful collection of remote sensing data are the diligent and quantifiable calibration of the instrument, the measurement and analysis of atmospheric properties, the proper application of radiative transfer algorithms that perform the atmospheric correction, and the ability to accurately collect or model surface properties such as reflectance and temperature.

Atmospheric correction generally requires at least three steps. First, the condition of the atmosphere is characterized and the column concentration of water present during the collection is identified. Second, the data is corrected based on this information and the reflectance is transformed. Third, during post-processing, any remaining artifacts are removed. The atmospheric correction process is optional and might not be required depending on the fidelity of the data required.

Modeling Target Interactions with Scattering by Arbitrarily Inclined Leaves (SAIL)

The interaction between light and vegetation—primarily leaves and grasses—is dominated by the surface characteristics, thickness, and vein structure of the vegetation. The primary function of a leaf is to capture the light and perform photosynthesis to create nutrients such as plant sugars. The natural evolution of the leaf has created a large surface area for the collection of light (Figure 2.7).

SAIL is a computational model to predict the spectral reflectance of uniform homogeneous vegetation canopies (Verhoef, 1984). SAIL is based on the radiative transfer theory to model the electromagnetic energy through different levels of foliage and canopies. The three sources of flux flow are the downward flux from direct radiation, the downward flux of diffuse radiation, and the upward flux of diffuse radiation. For the model to work properly, the reflectance and transmittance of the levels must be known or obtained. The scattering and extinction coefficients for canopies modeled in SAIL are calculated based on the fixed leaf inclination angle and a random leaf azimuth

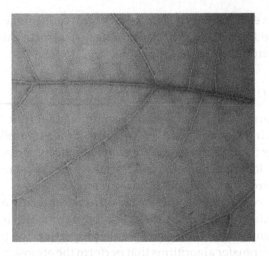

Figure 2.7 Leaf surface with vein structure. (Courtesy of Sam Davies.)

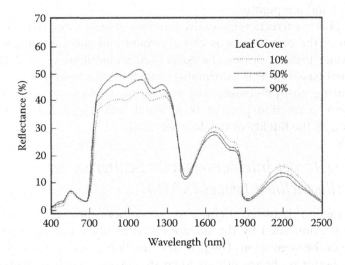

Figure 2.8 SAIL model simulations. (Hunt, Jr. and Parker Williams, 2006.)

distribution. Figure 2.8 shows SAIL model simulations of canopy spectral reflectance for various proportions of leafy spurge cover and grasses.

Other model parameters are the leaf area index (LAI), soil reflectance, diffuse skylight, and illumination and viewing angle. The LAI is defined as the one-sided area of leaves per unit area and has values from 0 to 16. Soil reflectance is a percentage less than 100 and is based on the albedo. The soil reflectance must be known for each spectral band used in the model. Diffuse skylight is dependent on the cloud cover and could range as high as 100% with completely overcast skies to as low as 10% on clear sunny days. Illumination and viewing angle are required because the leaf surface is non-Lambertian.

The two-component SAIL model (SAIL2) also takes into consideration optical and structural characteristics for leaves and stems, background reflectance, and the relative abundance of these components.

References

Hunt, Jr., E. R., A. E. Parker Williams. 2006. Detection of flowering leafy spurge with satellite multispectral imagery, *Rangeland Ecology Management*, 59(9), 494–499.

Ray, T. W. 1994. *Vegetation Remote Sensing FAQs*, Div. of Geological and Planetary Sciences, California Institute of Technology, www.yale.edu/ceo/Documentation/rsvegfaq.html.

Verhoef, W. 1984. Light scattering by leaf layers with application to canopy reflectance modeling: The SAIL model, *Remote Sensing of Environment*, 16, 125–141.

Weisstein, E. 2005. Solid angle, In: *MathWorld*, Wolfram Research, http://mathworld.wolfram.com/SolidAngle.html.

3

Imaging Spectrometers: Operational Considerations

The goal of this chapter is to provide the basic background information you will need to design and evaluate the operational aspects of hyperspectral remote sensing projects. If you are familiar with the components of a remote sensing mission, you will be in a better position to assess the utility, value, and accuracy of the data in hand or to be collected. This is a critical aspect of problem solving through remote sensing. This chapter will describe sensors and how they work, clarify common confusions and misconceptions, describe typical sensor system configurations, provide an example mission plan, and discuss workflow considerations.

Sensors

To understand hyperspectral imaging, first you need to understand its sensors and how they collect data, the number of bands and bandwidth, the type of platform from which the data is gathered, and the resolution of the data. In general, sensors gather data either passively or actively. Passive sensors collect and record electromagnetic energy that is reflected or emitted by surface features, typically through an optical lens. Examples include film or digital cameras and thermal infrared sensors, which detect emitted heat energy. Active sensors generate their own energy and then collect the signal that is reflected from the surface of the Earth. Examples of active sensors include RADAR (Radio Detection and Ranging) and LIDAR (Light Detection and Ranging).

For hyperspectral imagery, the data source includes ten or more bands of data. The bandwidth of the data typically ranges from 1 to 15 nanometers (a nanometer is one-billionth of a meter). In contrast, multispectral data typically consist of 3 to 7 bands of data with bandwidths ranging from 50 to 120 nanometers.

The platform from which the data is collected is either spaceborne or airborne. Spaceborne refers to satellite sensors, such as Landsat, Ikonos, QuickBird, ASTER (Advanced Spaceborne Thermal Emission and Reflection Radiometer), or Hyperion. Hyperion is a hyperspectral spaceborne sensor (using the strict definition of hyperspectral—number of bands and bandwidth). Airborne refers to fixed wing (airplane) or rotary (helicopter) platforms. Examples of airborne hyperspectral sensors include AISA, AVIRIS (Airborne Visual and Infra-Red Imaging Spectrometer), CASI, and HyMap.

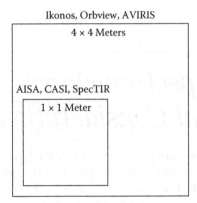

Figure 3.1 Spatial resolution.

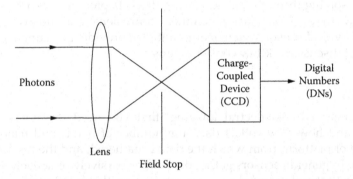

Figure 3.2 Electro-optical sensor.

Finally, resolution has two components, spatial and spectral. As explained in Chapter 1, spatial resolution (Figure 3.1) traditionally has been defined in the context of analog or film cameras as the smallest discernible spatial frequency. This frequency is measured in two ways: 1) in the laboratory using a pattern of points or bars with uniform spacing and measured in millimeters, and 2) outside the laboratory measuring the dimensions of resolvable features on the ground, expressed in coordinate units such as meters, feet, or inches. The laboratory-based method is typically used for camera calibration, while the ground measurements, known as ground-resolved distance (GRD), are reported to users for mapping purposes.

Digital imagery is created using an electro-optical camera or sensor. The basic components of an electro-optical sensor include the charge-coupled device (CCD) or chip, a lens, and a field stop (Figure 3.2).

The CCD is essentially a computer memory circuit composed of silicon detection elements (pixels) that are sensitive to photons, analogous to the silver halide-coated film in a traditional camera. The CCD accepts photons in the range of wavelengths that it is sensitive to (dynamic range) as a collection of electrons or a charge that represents the intensity and wavelength of the light received. An analog-to-digital converter changes the charge into an intensity value known as a digital number (DN). The lens provides focus and magnification of the photons, while the field stop or slit limits the amount of photons that pass through to the CCD.

Working together, the basic sensor components define the range of angles through which incident light or photons travel in reaching the CCD. This is termed the "field of view" (FOV). All features within the FOV are imaged, while those outside the FOV are not. Within the FOV at any given moment, one of the detection elements is exposed to light. The size of the detection element or pixel determines the range of incident angles of light received. This is called the "instantaneous field of view" (IFOV). From a spatial perspective, the IFOV delineates that portion of the ground that is being imaged. This is known as ground-sampled distance (GSD). Thus, in terms of resolution, the smallest resolvable feature would have the same extents as the GSD. Image geometry, platform motion, angle of illumination, and atmospheric conditions also affect resolution.

Although airborne multispectral and hyperspectral sensors typically have higher resolution, recently there has been a convergence of spatial resolution between satellite and airborne sensors. Spectral resolution, measured in bandwidths, remains the primary discriminator between multispectral and hyperspectral data, regardless of platform. One way to visualize bandwidth is to consider it as slices of, or the sampling interval along, a spectral curve. Multispectral satellite data has bandwidths ranging from 50 to 120 or more nanometers. This contrasts with common hyperspectral bandwidths, which range from 1 to 15 nanometers. Because hyperspectral sensors scan so many more bands than multispectral sensors, the spectral resolution is much greater (Figure 3.3).

An additional consideration is the location of the spectral bands within the electromagnetic continuum. Multispectral data can have gaps between the collected spectral bands, in contrast to hyperspectral data, which typically have many contiguous bands. Figure 3.4 contrasts the bandwidths of the AISA airborne and IKONOS spaceborne sensors. You can see that many AISA spectral bands will give greater spectral resolution than the four IKONOS bands.

The spectral resolution of a sensor is usually reported through two components, spectral sampling and full width at half maximum (FWHM). As described above, photons or light energy impinging on CCD detection elements, or pixels, yield an electron charge, which is converted into DNs along a response curve. Figure 3.5 provides an idealized pixel response curve that assumes a Gaussian shape. Spectral sampling refers to the interval at which DNs are collected or sampled along the response curve. FWHM refers to the detector response derived from exposure to a calibrated monochromatic source.

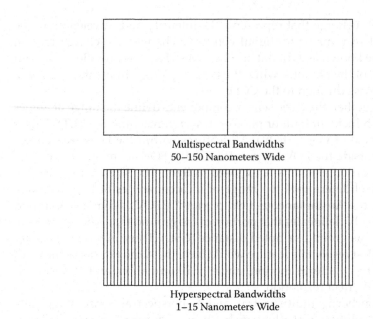

Multispectral Bandwidths
50–150 Nanometers Wide

Hyperspectral Bandwidths
1–15 Nanometers Wide

Figure 3.3 Number of multispectral bands compared to the number of hyper-spectral bands in the same area.

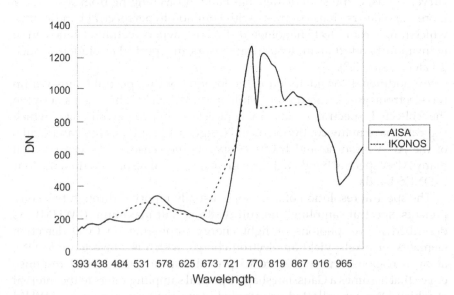

Figure 3.4 Spectral resolution.

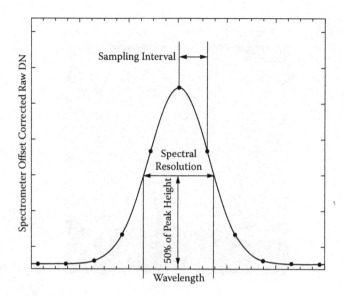

Figure 3.5   Full width at half maximum (FWHM).

The assumptions are that this response is Gaussian in nature, that the sampling interval defines the curve, and that FWHM provides a means to measure the width of this curve. It describes the channel bandwidth, reported in microns or nanometers, and specifies the spectral resolution. Analogous to spatial resolution, before a unique spectral response feature can be detected, the feature must be larger than the FWHM interval of the collection system.

Confusions and Misconceptions

The benefits of high spatial and spectral resolution come with a cost in complexity. Confusions refer to factors that complicate the processing and analysis of hyperspectral imagery. Some of these factors are illumination, mixing and proximity, and condition.

As described above, hyperspectral imaging sensors are optical devices that detect the response of surface features to incident light, or illumination. Illumination effects include the amount of light, angle of incidence, and atmospheric conditions. In terms of amount, the time of day is an important consideration in mission planning and data acquisition with the ideal time period being between 10:00 a.m. and 2:00 p.m. This provides for maximum illumination and a favorable angle of incidence, or sun angle. The usual objective is to minimize shadows and tonal variation, and as much as possible, light spillage, which is the reflection of light from one feature onto another.

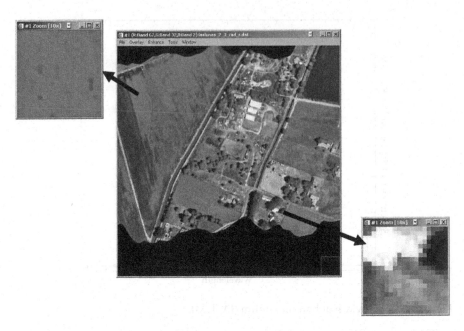

Figure 3.6   Mixing and proximity.

Atmospheric conditions such as cloud cover, moisture content, haze, and particulate levels affect illumination by absorbing, scattering, refracting, or reflecting light waves. This can result in obscuring some features or modifying their spectral response as detected by the hyperspectral sensor.

Mixing and proximity of surface features affect the ability of the sensor to detect and characterize the features of interest. Using vegetation as an example, considerations include whether a "patch" is dominated by one species (homogeneous) or many species are intermixed (heterogeneous), the morphology of the feature (e.g., size, growth form), phenology (e.g., deciduous or coniferous, annual or perennial), and number of features (e.g., dense or sparse). In addition, surrounding features can affect the spectral expression of a feature of interest. This can occur through shading or shadows, or reflection of radiation from adjacent features. Figure 3.6 shows an example of mixing and proximity. The zoom box on the left side of Figure 3.6 illustrates a relatively homogeneous patch of vegetation yielding a consistent spectral signature. The zoom box on the lower right illustrates an example of a building with a high emissivity roof that will affect the spectral signature of surrounding features.

Condition is another factor that modifies the spectral expression of a feature of interest. Condition includes the age and weathering of a material, moisture level, and health and vigor of the vegetation. Figure 3.7 shows an example of condition. The tonal variation visible on the rooftop in Figure 3.7 indicates differential weathering of the dark roof material, which will affect the characteristics of its spectral signature across the material.

Figure 3.7 Condition.

System Integration and Configuration

This section provides a general overview of the configuration of a typical air-borne hyperspectral sensor system. The configuration of a sensor installation varies by sensor and platform type but incorporates certain components in common. The term "system integration" as used in this context encompasses the major components of airborne sensor configurations, including optical sensors, inertial measurement units, airborne Global Positioning System (GPS) data collectors, and the flight management system. Additional considerations not covered here include physical details such as power requirements, power inverters, cabling, rack design and configuration, and weight management.

The various types of commercially available hyperspectral sensors can be distinguished in several ways, including spectral and spatial resolution, electronic design, and scanning geometry. In the context of system integration, scanning geometry provides a primary distinguishing characteristic and highlights other components of the sensor system. Most sensors use either a "whiskbroom" or "pushbroom" method to collect photon data for image creation.

The whiskbroom scanner uses a single detector and a rotating mirror that scans perpendicular to the direction of flight (crosstrack) to collect reflected light energy from a spot on the ground. The spatial extents of this spot are a function of the aircraft height above ground level (AGL) and the instantaneous field of view, as determined by the focal length of the sensor optics and internal design considerations. Whiskbrooms typically scan a crosstrack arc ranging from 90 to 120 degrees in extent, with image pixel size increasing at the extremes of the arc. Complexities in image geometry are introduced by the crosstrack-scanning pattern, the forward motion of the aircraft, the variability in image pixel size, and the optical distortion.

The pushbroom scanner consists of a linear array of detectors that record a single line of the ground in the crosstrack direction. The number and dimensions of the detectors associated with the CCD comprise a line, often referred to as a frame. Each individual detector records spatial coordinates (x, y) and an entire spectral curve (z) as defined by the dynamic range of the CCD. The spatial extents of the area imaged on the ground are a function of the detector size, aircraft height (AGL), focal length of the sensor optics, and internal design considerations. The forward motion of the aircraft, optical distortion effects, and detector integration time, which can cause smear, are the primary sources of image geometry complexity. One other consideration is the requirement for detector calibration and normalization of spectral data.

A sensor component that is not found in every configuration, but serves a valuable purpose, is some form of downwelling irradiance sensor (DIS). A DIS is mounted on the outside of the aircraft and pointed skyward to measure the incoming (or downwelling) light energy as it passes through the atmosphere and strikes the Earth's surface. Light energy reflected from the Earth's surface is collected by the hyperspectral sensor in radiance units, or DNs. This can be termed "at-sensor radiance" for the spectral dynamic range of the sensor.

Each pixel contains all wavelengths determined by the dynamic range of the sensor. Dividing at-sensor radiance for each wavelength within each pixel by the corresponding downwelling radiance for that wavelength yields the "at-sensor" reflectance. In other words, at-sensor reflectance, calculated using this method, is a spectral value that accounts for ambient light energy by normalizing detected radiance values based on downwelling radiance values.

Two integral components are necessary for geometric correction of the output data, the inertial measurement unit (IMU) and an airborne global positioning system GPS (AGPS). The IMU, which is composed of accelerometers, gyros, and a data recorder, is physically attached to the sensor and collects information on the attitude and orientation of the sensor. The AGPS collects aircraft position data at regular intervals using radio signals received from a constellation of positional satellites orbiting the Earth.

Postprocessssing software combines the IMU and AGPS data streams to calculate aircraft geographic position, velocity, and heading, as well as the roll and pitch of the aircraft for each interval. This information is used to correct the geometry of the data and relate each pixel in the output imagery to a location on the ground (Mostafa et al., 2001). The data collector controls the actions and settings of the sensor and enables monitoring and storage of the data stream.

Flight management system (FMS) configurations vary by system and application. Of note in the context of airborne data collection is a component that stores and displays the programmed flight lines and current aircraft position and attitude. The flightlines are developed as part of the mission planning activity and typically input to the flight management system in the form of computer-aided design (CAD) files. (For additional information about system integration and configuration, see Mostafa et al., 2001.)

Mission Planning

Data-gathering missions are expensive and time-consuming, so planning and organizing as much as possible about the mission is crucial for obtaining the highest quality data possible. Obviously, some components cannot be controlled, the most obvious being the weather conditions and amount of sun available for the chosen day. Other components, however, must be defined before the mission day: time of day for the flight, altitude, sidelap, flightline orientation, area of interest (AOI), number of flightlines, horizontal accuracy, coincident field work, and boresight flight.

Factors relating to the actual aircraft include the required altitude; flightline orientation, which is the direction the airplane flies (e.g., north to south); number of flightlines, which is the number of times the aircraft flies over the designated area to collect data; horizontal accuracy; and boresight flight.

Factors relating more to the sensor and collection of data include the AOI, sidelap, coincident field work, and archival data. Obviously, the pilot needs to know the location of the specific AOI, which is the point, line, or polygon chosen as the image area to be used in the mission. The scanner itself will have to be calibrated for the amount of sidelap. In a block of mapped data consisting of a number of parallel bandwidths, the sidelap (measured vertically) is the overlap of the edges of the bands (Figure 3.8).

Another step of the data-gathering mission is verifying the accuracy of the new data. The accuracy of the gathered data can be determined by comparing the data gathered from the airborne mission to data collected at the same time during coincident field work or to archival data.

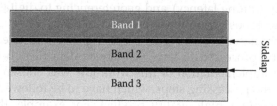

Figure 3.8 Sidelap.

Workflow

Figure 3.9 shows an example of a hyperspectral data processing workflow. In any workflow diagram, the number and particular order of tasks will vary by the type of sensor, type of software, analyst experience, and desired product. Thus, some workflow charts will necessarily be either more or less complex.

The workflow in Figure 3.9 is divided into three major tasks: data file conversion (C), radiometric correction (R), and geometric correction (G). Each step in the process is described below to illustrate the necessary tasks associated with each component.

During the first task, data are collected in three streams: dark frame data that provide sensor noise information, raw image and associated data (band wavelengths and widths), and geolocation data (GPS). The dark frame consists of a series of exposures taken with the lens shutter closed. This information is used to model the detector chip noise and light energy thresholds. The dark data are saved in a separate file. The raw hyperspectral data consists of intensity values for each wavelength and pixel, a row and column value for relative location, and a time stamp denoting when each image frame was collected. The image, band, and GPS data are compiled in a single flat file. These file formats are then converted during step C.1 to a format that is compatible with the image processing software modules. During this process, the data components are extracted and written to separate files. These files include the intensity or DN file; a band file (BND), which describes the spectral range of the DNs; a navigation file (NAV), which contains the data frame acquisition time; and a dark frame, which has noise values relatively referenced by row and column.

The second task, radiometric correction, involves various steps to model sensor noise, atmospheric conditions, and illumination effects. A number of correction factors are applied as shown in the workflow diagram. Although much of this effort is standardized, the person processing the data will decide which procedures and implementations are most suitable to use for each application.

Finally, during the third task, geometric correction is used to address issues of geolocation in both the interim and final image products. The procedure depicted in Figure 3.9 uses airborne GPS and IMU data to achieve geolocation. This data is supplemented by boresight information (attitude and orientation) derived from a "patch test," GPS and sensor timing synchronization (timing latency), and georeferencing to digital orthophotos collected simultaneously with the hyperspectral data.

The workflow described in Figure 3.9 is highly interactive. In addition, all data must be processed to level G.6 (create data tiles) before analysis can begin. The implementation of this workflow was chosen to minimize redundancy in the processing steps, which have to be followed in sequence to assure an accurate, high-quality product. For example, the crosstrack (R.3: xtrack) correction removes the differential illumination effect of the sun

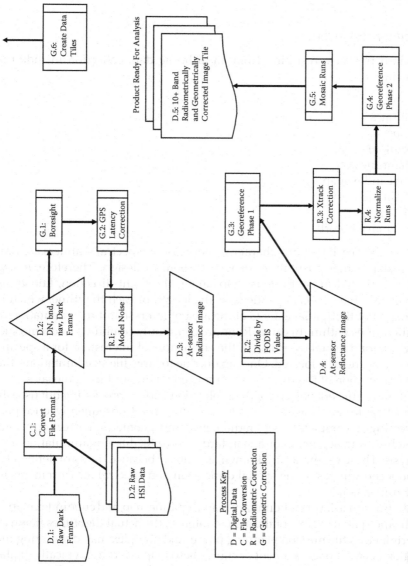

Figure 3.9 Hyperspectral data processing workflow.

angle for each flightline. This procedure requires that the entire flightline (or run) be processed to reflectance before it can be applied. If only a portion of the run were used, the correction coefficients would yield a nonstandard result, leading to confusion and error in feature classification based on the spectral signature.

Standard Products

Standard products resulting from an airborne data collection include the following:

Runs
Mosaics
Cubes
Boundary map
Flightline map
Cube index map
Flight log
Metadata
Summary report

As described earlier in this chapter, airborne hyperspectral scanners are predominately either pushbroom or whiskbroom in design. Therefore, image data is collected in continuous strips along the flight direction, known as "runs." Run file size is dependent on the length of each flightline, which is a function of the spatial extents of the area of interest and the orientation of the flightlines within this area. Runs are commonly combined into a mosaic, allowing for a visual overview of the entire area of interest, or for a specific area or feature to be located. Once again, due to large file size limitations, the native resolution of the runs is often reduced when producing mosaics.

Runs are efficient objects for data collection and preprocessing but usually impractical for analysis due to their large file size. Consequently, once preprocessing is completed (e.g., radiometric and geometric correction), runs are subset into segments of a standard size to facilitate manipulation and analysis. These segments are known as "cubes" because they are defined by x and y coordinates in the spatial dimension and spectral measurements in the z dimension.

The three map layers—boundary map, flightline map, and cube index map—define the spatial extents of the area of interest, the actual flightlines flown as determined in the mission planning phase, and an index map delineating the location of each image segment or cube. These map layers are typically digital in CAD (dxf, dwg, or dgn file extensions) and/or GIS (Geographic Information System) software (shp or map file extensions) formats.

The mission flight log is an invaluable source of useful information, aiding in processing and analyzing data, as well as providing the basis for project metadata. Log formats vary, but at a minimum should include standard, mission-specific, and flightline information. Standard information items are date, time of flight, aircraft, pilot, operator, and equipment (including model and serial number). Mission-specific information includes airport name and identifier, elevation, air temperature, wind speed, and cloud cover. Flight-line information records flight direction, airspeed, altitude (above ground level), flightline number, integration time, and operator notes (e.g., changes in sensor settings, atmospheric conditions, and data collection issues).

Finally, the airborne data collection includes a summary report that highlights the significant findings and conclusions of the mission.

Reference

Mostafa, M., J. Hutton, and B. Reid. 2001. GPS/IMU products—the Applanix approach, In: Fritsch/Spiller (eds.), *Photogrammetric Week 2001*, Wichmann Verlag, Heidelberg, Germany, pp. 63–83.

4

Hyperspectral Remote Sensing and the Atmosphere

Atmospheric Interactions

Because hyperspectral data is collected some distance above the target on the ground, the reflected solar illumination must travel through the atmosphere. Characteristics of the atmosphere can have a profound effect on the incoming solar energy recorded by the hyperspectral sensor. The way light interacts with the atmosphere depends on many factors: the types of particulates and gases, the amount of atmospheric reflection, atmospheric absorption, and atmospheric scattering. The atmosphere is a complex mixture of particulates and gases. Particulates are small particles usually less than 20 μm in diameter. Particulates greater than 20 μm are not likely to stay in the atmosphere for long durations and tend to settle to the ground rather quickly.

Amount of Atmospheric Reflection

Specular reflection is the perfect mirrorlike reflection of light off a smooth surface at the corresponding angle of incidence (Figure 4.1).

Amount of Atmospheric Absorption

Atmospheric absorption is the removal of energy from solar irradiance by conversion of the electromagnetic energy to another form, usually thermal energy. Atmospheric absorption occurs when a photon induces a molecular vibration, rotation, or electron orbital transition to an alternate energy state. The photon is absorbed by the constituent molecules of the atmosphere, and only photons with specific energy levels can be absorbed.

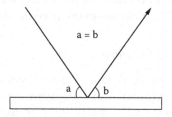

Figure 4.1   Basic reflectance off a smooth surface.

Table 4.1　Principal Molecular Absorption Lines in the
Earth Atmosphere

Wavelength (μm)	Molecule	Wavelength (μm)	Molecule
0.26	O_3	3.9	N_2O
0.60	O_3	4.3	CO_2
0.69	O_2	4.5	N_2O
0.72	H_2O	4.8	O_3
0.76	O_2	4.9	CO_2
0.82	H_2O	6.0	H_2O
0.93	H_2O	6.6	H_2O
1.12	H_2O	7.7	N_2O
1.25	O_2	7.7	CH_4
1.37	H_2O	9.4	CO_2
1.85	H_2O	9.6	O3
1.95	CO_2	10.4	CO_2
2.0	CO_2	13.7	O_3
2.1	CO_2	14.3	O_3
2.6	H_2O	15	CO_2
2.7	CO_2		

Source:　Rees, W.G., *Physical Principles of Remote Sensing*, 2nd ed.,
Cambridge University Press, 2001. With permission.

Molecular Absorption

Molecules can absorb electromagnetic radiation in three ways: electronic transitions, vibration, and rotation. Electronic transition requires the greatest amount of energy and involves the promotion of electrons to higher energy levels. Vibration involves the molecular bond between atoms and models it as a spring. Rotation can be considered in the context of a simple diatomic molecule of two atoms. The two atoms can rotate about their center of mass. Molecular absorption can be complicated by a combination of mechanisms occurring simultaneously. For example, the energy level of a molecule can be described by both rotation and vibration. Table 4.1 summarizes the main absorption lines in the Earth's atmosphere (Rees, 2001).

Amount of Atmospheric Scattering

The surface area of the particulate and gases interacts with the light and creates scattering or a redirection of the electromagnetic radiation (EMR), which can cause a change in the distribution of the EMR. The interaction of the light with the particulate or atmospheric gases that do not change the properties of the light is called "elastic scattering." The dynamics of the scattering are based on the particulate or gas surface geometry and characteristics. Atmospheric scattering results in fewer photons reaching the collection optic sensors and fewer photons available for the sensors on the focal plane array or other detector elements.

Chandrasekhar (1960) and Van de Hulst (1981) present a treatment of the complex details of scattering theory and the related radiation propagation (Schott, 1997). The three basic types of scattering are classified as Rayleigh, Mie, and non-selective scattering. Rayleigh scattering occurs when the EMR interacts with the minute particles or molecules that are the components of the atmosphere, primarily when the particles are much smaller than the wavelength of the incident flux. Mie scattering results when the wavelength of the incident EMR is approximately equal to the size of atmospheric particles, such as aerosols, dust particles, fossil fuel combustion products, and suspended sea salts. Nonselective scattering occurs when the suspended atmospheric particles are very large with respect to the incident EMR; particles such as water droplets and ice crystals can cause nonselective scattering (Schott, 1997).

Rayleigh Scattering

In attempting to explain the blue color of the sky, Lord Rayleigh (1871) first described the phenomenon of scattering. He characterized the fractional amount of energy scattered into a solid angle at an angle θ from the propagation direction per unit length of transit in the medium. Rayleigh scattering involves an inverse dependence with wavelength to the fourth power and an inverse dependence on number density.

Rayleigh scattering, which generally occurs at high altitudes in the upper atmosphere, is also referred to as clear air scattering because of the minor amounts of particulates that are available for scattering. The primary components for the scattering at these altitudes are atmospheric gases including oxygen and nitrogen. Rayleigh scattering is dependent on the wavelength of the light, and the scattering increases as the wavelength becomes shorter. Another property of Rayleigh scattering is that the particles have a smaller diameter than that of the incident light wavelength.

The size of a Rayleigh scattering particle is defined as:

$$x = \frac{2\pi r}{\lambda}$$

where x is the size of the particle, which is much less than the incident light wavelength.

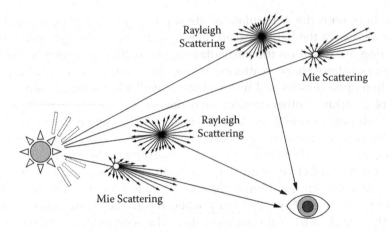

Figure 4.2   Basic principles of Rayleigh and Mie scattering.

The wavelength dependence explains how the sky is blue because more short wavelength energy is scattered from the incoming solar flux. Similarly, the sky is red at sunrise and sunset because the remaining solar flux will contain proportionately greater amounts of longer wavelengths as the shorter wavelengths are removed by scattering (Schott, 1997).

Mie Scattering

When you look at the sky, it looks bluer the farther you look from the sun. The almost white scattering near the sun can be attributed to Mie scattering, which is the second basic type of scattering. Mie scattering occurs closer to the ground and up to altitudes of 3 miles (5 km). Mie scattering is caused by the interaction of light with larger particulates such as dust, pollen, smoke, and water droplets. The particulates generally have a diameter equal to the incident light wavelength. The interaction is not as wavelength dependent as Rayleigh scattering, and the primary impact is to light in the visible portion of the spectrum.

As shown in Figure 4.2, Mie scattering is mostly forward scattered, unlike Rayleigh scattering, which is symmetric with approximately equal amounts of forward and backscatter. Also, Mie scattering is not as dependent on wavelength as Rayleigh scattering.

Nonselective Scattering

The third basic type of scattering is nonselective scattering, which occurs at the levels closest to the ground where the particles are usually much larger in diameter than the wavelength of the incident light. Nonselective scattering does not depend on the wavelength and the scattering occurs uniformly in all directions. This type of scattering usually involves large dust particles,

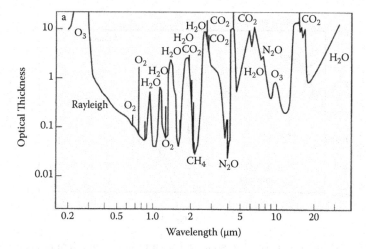

Figure 4.3 Total zenith optical thickness of the standard atmosphere for the ultra-violet, optical, and infrared region (modified from Rees, 2001).

water droplets, ice, and hail, and is the primary factor in the haze experienced in urban environments.

Figure 4.3 shows the optical thickness as a result of the scattering and absorption of electromagnetic radiation propagating vertically through the atmosphere. Figure 4.3 is based on "standard" atmospheric conditions and only shows general spectral detail. The near-IR region shows the many absorption features related to molecular transitions. Since Figure 4.3 assumes vertically incoming electromagnetic radiation, an oblique path would result in greater optical thickness (Rees, 2001).

Atmospheric Transmission

Figure 4.4 shows atmospheric transmission as a function of wavelength for a standard atmosphere. This figure also shows the solar exoatmospheric spectral irradiance in $wm^{-2}\mu m^{-1}$, and the radiant exitance from a 300 K blackbody ($wm^{-2}\mu m^{-1}$). There are several orders of magnitude more flux from the sun than the emittance from the 300 K blackbody for the visible through the short-wave infrared.

Atmospheric constituents have unique spectral absorption characteristics and affect the incoming solar irradiance. Figure 4.5 shows the absorption spectra of various atmospheric constituents for a single pass through the U.S. standard atmosphere for a 45-degree solar illumination path. The bottom curve in Figure 4.5 shows the actual atmospheric transmission as a result of combining the individual spectra for each of the atmospheric constituents. Figure 4.6 shows the exoatmospheric spectral irradiance combined with the atmospheric transmission spectra (Schott, 1997).

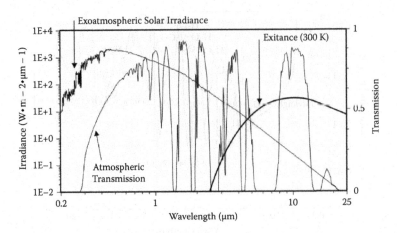

Figure 4.4 Comparison of atmospheric transmission, solar irradiance, and self-emission spectra (Schott, 1997).

References

Berk, A., L. S. Bernstein, and D. C. Robertson, 1989. "MODTRAN: A moderate resolution model for LOWTRAN 7," Technical Report GL-TR-89-0122, Air Force Geophysics Laboratory, Bedford, Burlington, MA.

Chandrasekhar, S., 1960. *Radiative Transfer*. Dover, Mineola, NY.

Lord Rayleigh (J. W. Stratt), 1871. *Philosophical Magazine*, Vol. 41, pp. 107–120, 274–279.

Rees, W. G., 2001. *Physical Principles of Remote Sensing*, 2nd ed. Cambridge University Press.

Schott, J. R., 1997. *Remote Sensing: The Image Chain Approach*, Oxford Series in Optical and Imaging Sciences. Oxford University, June 1997.

Van de Hulst, H. C., 1981. Evaluating the light from the sun. *Optical Spectra*, 6(3), 32–35.

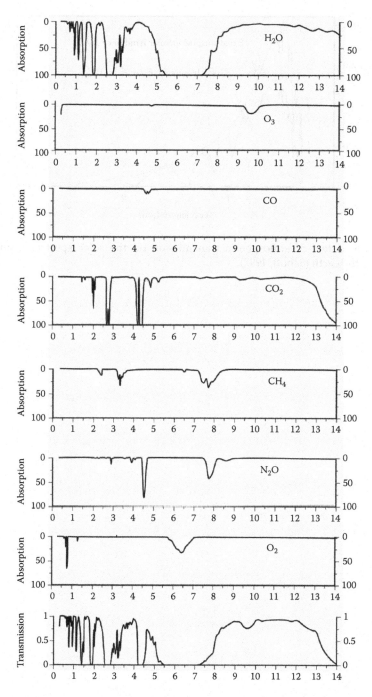

Figure 4.5 Absorption spectra of various atmospheric constituents and overall atmospheric transmission as derived from MODTAN (Schott, 1997).

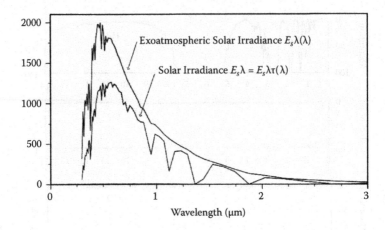

Figure 4.6 Effect of atmospheric transmission on the solar spectral irradiance reaching the earth (Schott, 1997).

5

Information Extraction from Optical Image Data

Data vs. Information

Data are a collection of observations, or raw, unprocessed values or sets of values. Information is data within a given context. Without the context, the data are usually meaningless, as depicted in Table 5.1.

Once you put the information into meaningful context, you can use it to make decisions. The transfer of information to the person who needs it, such as a data analyst or policy maker, can increase the ability of that person to make better decisions. A data value may contain information in a specific context. For example, when you want to calculate blackbody radiation, Planck's constant is the information you need to derive the answer. Otherwise, that number is just one data element in the world of physics.

The development of computers has made it possible to process large amounts of data into information. Even the fastest computers, however, have difficulty processing all the data gathered by the larger focal plane arrays that are built into hyperspectral systems with greater spectral resolution. One of the disadvantages of these hyperspectral sensors is the time it takes to convert collected observations into information that can be used for intelligence, knowledge, or decision making.

Probably the most important characteristic of good information is its relevance to the problem. Information is usually considered relevant if it helps improve the decision-making process. If the information is not specific to your problem set, it is irrelevant. For example, information about the mineral content of surface soils in Nevada is only useful if you plan on searching for minerals in Nevada. Otherwise, the information is not relevant.

Timeliness and accuracy are also strong considerations for the value of the information. Timeliness of data or information is directly related to the gap

Table 5.1 Data vs. Information

Data Value	Information
300,000,000	Population of the United States in 2006 census; speed of light measured in meters per second
500	Time in seconds it takes for light to travel from the sun to Earth
2π	Circumference of a unit circle

between the occurrence of the event to the transfer of information to the user. A system is considered "real time" when the gap between data collection and product development (such as target detection) is very short.

Accuracy is the comparison of the data to actual events. Many times, a data authentication process is used to determine the validity of the data collected.

In hyperspectral remote sensing, the ability to derive information from spectral data is the key to any successful collection. The vast amount of spectral data must be culled to define the spectral signature of interest for the material under consideration. In spectral terms, the pure spectral signature of a feature is called an endmember. One method of collecting pure endmembers is from a laboratory spectroradiometer that is focused on a single surface or material. These signatures are then used in the spectral sensor, and detection algorithms are used to define and refine the spectral scene collected so a material or materials with similar characteristics can be defined. However, when the material of interest is not available for laboratory measurements, it must be defined within the spectral scene collected.

Classification Style/Intent

How people visualize data, information, and the world around them is an important resource in many areas of analysis, research and development, and theoretical studies. Although helpful in evaluating complex tasks and designs as well as imagery data, visualization and classification are not widely used in routine data analysis in many fields because the software with intuitive graphical user interfaces that allows the representation of ordinary data sets has not been developed. Classification and visualization software requires complex algorithms that are usually not cost effective for the evaluation of ordinary tasks and data sets.

An image analyst determines the classification approach and decides between using spectral classes or information classes. A cluster of pixels with nearly identical spectral characteristics is considered part of a spectral class. An analyst uses an information class, such as pine trees, orange trees, or gravel, when trying to identify specific items or groups within an image. The primary goal of an image analyst is to try to match the spectral class to an information class. For example, in Figure 5.1, each set of five pixels has been identified as a spectral class. If the analyst knows that this spectral class has the characteristics of a pine tree, these spectral classes are assigned to the pine tree information class.

Supervised and Unsupervised Classification

Once the analyst has decided to use spectral or information classes, the classification process can be either supervised or unsupervised. A supervised classification is based on detection algorithms using pixels from known reference samples, usually located within a scene, as a basis for comparison

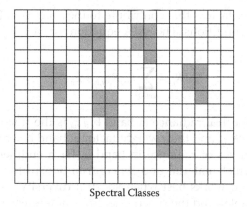

Spectral Classes

Figure 5.1 Spectral classes.

to other pixels from objects in the same scene. For example, if the analyst knows one specific area is a gravel road, then all other areas with the same detection algorithm will also be a gravel road. Therefore, in supervised classification, the analyst usually starts with known information classes that are then used to define representative spectral classes that closely match the reference samples.

Unsupervised classification is basically the opposite of supervised classification. Pixels in an image are grouped into spectral classes based solely on the information in the data compared to signature libraries or other known information classes.

Once the data set is minimized to a size that can be processed and exploited, image classification can be used to assign spatial and spectral information into various "themes," commonly referred to as "thematic maps." These themes are used to depict natural items such as soil or other geological features, vegetation, water, or man-made objects such as structures or vehicles.

Feature Extraction

Even though current remote sensors and data collection systems often create extremely large data sets that are difficult to work with, the information contained in these data sets can be valuable. As a result, software has been developed to aid with the visualization and classification. One option for the image classification process is feature extraction. This option reduces the spectral or spatial characteristics with spectral transformations or spatial filters so data sets can be easily processed and exploited. Feature extraction can also be accomplished by selecting a subset of bands based on the characteristics of certain items of interest.

One method for isolating spectral features is called Spectral Mixture Analysis (SMA). SMA is a structured approach that addresses the mixed-pixel problem and other factors that contribute to the image quality, such as calibration and light conditions. The SMA equation for each band is

$$R_b = \sum_{em-1}^{N_k} F_{em} R_{em,b} + E_b$$

where R_b is the spectral radiance at band b, F_{em} is the fraction coefficient of each endmember, R_{em}, and their weight factor at band b, while E_b is the error for any other sources of radiance in band b. Each endmember is selected based on its distinct material and its contribution to the overall spectral scene. This method works best when spectral diversity and content of the scene are not complex and the spectral features of interest are very minor in the scene.

Other methods of feature extraction include the first difference Partial Least Squares (PLS) regression, which uses a Singular Value Decomposition (SVD) of the entire spectrum within the scene, and Hierarchical Foreground/ Background Analysis (HFBA), which divides the spectral scene into two groups, foreground and background, that contain the spectral signature of the feature or features of interest.

Algorithm Development

An algorithm is a step-by-step procedure that terminates after a finite number of steps. An algorithm is different from a computer program. An algorithm, which can be written in any language, including English, is more like the reason for developing the program. The program has to be written in a particular programming language, but the steps of the algorithm are what perform the task. The steps must be unambiguous so that carrying out the steps will accomplish the assigned task.

With all the remote sensors collecting spectral data for government and commercial uses, detection algorithms have become crucial for the data to be able to be used by consumers throughout the world. Being able to process the hyperspectral data is as important as the actual sensor system. Many scientists and researchers have worked on developing and implementing robust detection and identification algorithms that will make the hyperspectral data useful for both commercial and military markets.

Most detection algorithms require access to the spectral attributes in a spectral signature library. Spectral libraries contain reference spectra either measured or simulated from field and laboratory collections of reflectance or radiance data. After pixel spectra are collected, the new spectra are compared to the pixel spectra in the spectral library for detection and identification.

Spectral analysis methods usually use image analysis algorithms to compare pixel spectra with a reference spectrum (often called a "target spectrum" or "endmember"). The most commonly used algorithms for hyperspectral and multispectral image processing are whole-pixel analysis, spectral angle mapper (SAM), spectral feature fitting, sub-pixel analysis, complete linear spectral unmixing, and matched filtering. These algorithms are included in commercially available visualization software packages such as Environment for Visualizing Images (ENVI™) from ITT and Earth Resources Data Analysis System (ERDAS) developed by Leica.

Whole-Pixel Analysis

The whole-pixel analysis method is used to determine if one or more target materials are abundant within each pixel. The spectral similarity of the target pixel is then compared with materials in a reference library. Whole-pixel analysis is often used against standard supervised classifiers or the SAM and spectral feature fitting analysis methods.

Spectral Angle Mapper (SAM)

SAM considers every pixel in the scene and evaluates the similarity of the spectra to repress the influence of the shading, which accentuates the characteristics of reflectance. The image spectrum is then assigned a correlation factor between 0 (low correlation) and 1 (high correlation) and compared to a spectral library or endmember. With SAM, the data are converted to apparent reflectance, which is the true reflectance with gain coefficients that are defined by terrain and lighting conditions. A scatter plot of pixel values from two bands of a spectral image can be generated to visualize the spectral components. The plot in Figure 5.2 shows the pixel spectra and target spectra as points. Figure 5.3 shows a SAM plot of various materials.

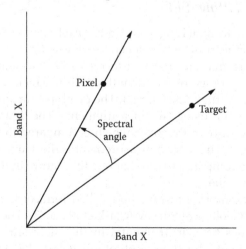

Figure 5.2 Pixel spectra and target spectra as points.

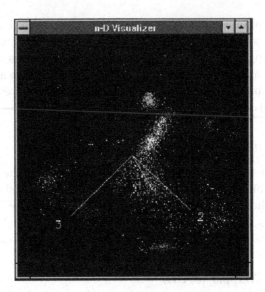

Figure 5.3 Spectral Angle Mapper (SAM) plot of various materials (courtesy of ITT).

The spectral angle is the angle between any two vectors originating from a common origin. The magnitude of the angle indicates the similarity or dissimilarity of the materials—a smaller angle correlates to a more similar spectral signature. This method is relatively insensitive to changes in illumination on the target material because changes in light will impact the magnitude but not the direction of the vector. A poorly illuminated target will cause the points to be plotted closer to the origin (Clark, 1999).

Spectral Feature Fitting (SFF)

Another approach to matching target and pixel spectra is by examining specific absorption features in the spectra. SFF is a detection algorithm that uses image spectra that are matched to reference endmembers. With SFF, the user specifies a range of wavelengths within which a unique absorption feature exists for the chosen target. The pixel spectra are then compared to the target spectrum using two measurements. The first measurement is the depth of the feature in the pixel, which is compared to the depth of the feature in the target. The second measurement is the shape of the feature in the pixel, which is compared to the shape of the feature in the target using a least-squares technique.

An advanced example of this method, called Tetracorder, has been developed by the U.S. Geological Survey (Clark et al., 2003). The U.S. Geological Survey, Denver, has been instrumental in the successful implementation of variations of the SFF for their applications (Clark, 1999) as illustrated in Figure 5.4.

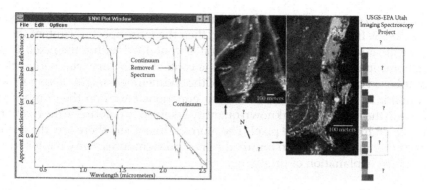

Figure 5.4 SFF analysis with and without continuum (left) and application to mineral detection and identification (right) (Clark, 1999, courtesy of USGS).

Tetracorder compares reference spectra with unknown spectra after the continuum has been removed. A continuum is a mathematical function used to isolate a particular absorption feature for analysis (Clark and Roush, 1984). The reference spectra are selected from a signature library or the collected imagery. The reference spectra or endmembers are then adjusted by subtracting the continuum-removed spectra. Then the reference spectra are scaled to match the unknown spectrum. The greater the scale factor, the less correlation there is between the signatures. A statistical analysis of least-squares fit between the curves is then performed, which results in a root means square that can be used to further correlate the statistical significance of the spectral match.

Sub-Pixel Analysis

The sub-pixel analysis method is a very powerful detection algorithm that can be used to calculate the quantity of target materials in each pixel of an image. Sub-pixel analysis can detect quantities of a target that are much smaller than the pixel size itself. In cases of good spectral contrast between a target and its background, sub-pixel analysis has detected targets covering as little as 13% of the pixel.

The more contrast the spectra have, the better these algorithms are in discriminating the spectral signature. Targets that fill several pixels also have greater statistical significance for an increased probability of detection. However, when the object of interest is significantly small, these algorithms can lead to initial detection and the possibility of further identification. The disadvantage of sub-pixel analysis is the number of false detections that occur when precise spectral measurements are attempted with a limited amount of information.

Complete Linear Spectral Unmixing

Complete linear spectral unmixing is a method for performing sub-pixel detection based on the theory that each pixel contains a linear combination of reflectance spectra from all the endmembers within the pixel (Adams et al., 1986). If the relationship is linear, then the more spectral content there is, the greater the amount of that material is contained within the pixel. This relationship can be used to calculate the content of the material within the scene based on the spectral signature of the endmember sought. The spectral unmixing is achieved by a set of linear equations based on the number of bands in the image. The unknown quantity is solved to determine the fraction of each endmember in the pixel. The more endmembers there are, the more bands that must be used to unmix the spectral signatures. See Chapter 6 for a detailed explanation of unmixing.

Matched Filtering

Along with complete linear spectral unmixing, matched filtering is another method for performing sub-pixel detection. Matched filtering is a specialized methodology for spectral unmixing that only uses specifically defined endmembers for the spectral maps. This varies from the complete linear unmixing routine, which requires all endmembers in the pixel to be defined. This method is commonly referred to as partial unmixing.

Matched filters are used to find spectral signatures for materials that constitute a relatively minor portion of the spectral content within the image. Matched filtering got its name because it filters the spectral scene to provide good matches to the endmember of interest and suppresses all other signatures. All other signatures are considered background.

This technique requires significantly less computational analysis and, therefore, can provide a rapid means to analyze, display, and exploit the imagery. As with other linear unmixing methodologies, false detections can be present in the final data. ENVI uses a tool to calculate how feasible the solution is based on noise and other image statistics to determine the correlation between the matched filtering result and what would be expected from a mixture of the target and the background under those conditions.

Match filtering was used on airborne hyperspectral imagery immediately following the Space Shuttle Columbia tragedy in Texas on February 1, 2003. It was critical to find as many space shuttle fragments as possible before the material degraded due to exposure to environmental elements.

As a result, space shuttle fragments as small as 10 sq. in. were detected using remote hyperspectral sensing and match filtering algorithms. These fragments proved critical for the space shuttle recovery effort and crash investigation, and demonstrated the value of remote hyperspectral sensing under time-sensitive conditions.

Information Contained in an Image

Spectral data is essentially high-resolution imagery that contains spatial and spectral information. Due to the quality of the GSD (ground-sampled distance)

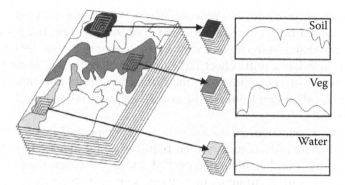

Figure 5.5   Interaction of reflected light with surface materials.

of many hyperspectral systems, single-band data is digital imagery that is geospatially registered. That means that every pixel has an associated x, y coordinate that is usually resolved from an inertial navigation system (INS), with adjustments for aircraft roll, pitch, and yaw.

The strength of hyperspectral imagery is the number of contiguous bands used to compile each scene. The reflected light from the surface interacts in a unique way for soil, vegetation, and water, and even slight variations of those materials are quantified and characterized through hyperspectral imagery (Figure 5.5).

The reflected light used by hyperspectral sensors is governed by the laws of physics. To maintain thermodynamic equilibrium, the sun does not lose gain of energy as it interacts with the vacuum of space, the atmosphere, and the Earth's surface. The sun accounts for most of the solar energy, which is mainly concentrated between 0.4 to 3.0 μm (Arya, 2001).

The fraction of solar energy that is reflected back into the atmosphere is called "albedo." When an object reflects most of the light that hits it, it looks bright and it has a high albedo. When an object absorbs most of the light that hits it, it looks dark and has a low albedo. The albedo of a perfectly white object is 1; the albedo of a perfectly black object is 0.

The albedo value depends on the interaction of the surface areas of reflective materials. Reflectance is usually considered from a single incidence angle because light usually reflects off the surface of an object at the same angle as the incident light. Surface geometry can change a portion of the incident light. Albedo takes into consideration reflectance at all angles and can be ascribed to the bidirectional reflectance distribution function (BRDF).

Most albedo values are derived from laboratory measurements because they are dependent on the portion of the electromagnetic spectrum under consideration from remote sensing. The mass median diameter (MMD) and other estimates of grain size and shape are commonly used for BRDF models. But as expected, estimations and simplifications can lead to a discrepancy in simulated and measured data.

The interaction of light as it reflects off a surface creates a spectral response that is detected by the sensor and then analyzed by detection algorithms. Any surface that is in the field of view of the sensor that does not have black-body characteristics will reflect light in a wavelength that is captured by the remote sensor. Depending on the diffraction optics and notched filters (sometimes called band filters), the spectral signatures will be resolved as spectral bands.

In many applications, the objective of hyperspectral image analysis is to detect and identify objects, which requires a high degree of confidence. Increasing the number of hyperspectral bands, however, does not necessarily lead to more accurate identification. So even though the vast amount of information is contained within the spectral signature, the selection of bands to extract the required information is critical. The use of too many bands can have a "diminishing returns" effect or even a negative impact on the ability to exploit the spectral scene. Because considering all the possible sets of bands for a single evaluation would be extremely time consuming, analytical methods were developed to aid with band selection (Price, 1994).

Concept of a Hyperspectral Cube

A spectral cube is a three-dimensional array containing spatial (image) information on the x and y axes and spectral information on the z axis (Figure 5.6). Individual spectra, spectral maps, and full spectral cubes can be created from a single spectral cube. Spectral cubes display different stratified, graphical thicknesses in a three-dimensional perspective that can be a useful tool for overall image analysis. The magnitude of the spectral signature is created by mapping a color to the intensity of the spectral response at different wavelengths at a given spatial area. Because spectral cubes are generally created over larger areas than the focal plane array can collect in one frame, the second dimension of the spatial image is

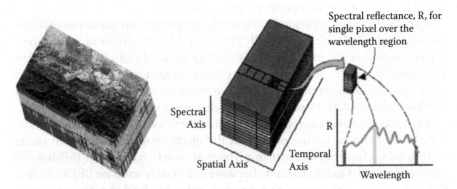

Figure 5.6 Spectral cube and the fundamental components.

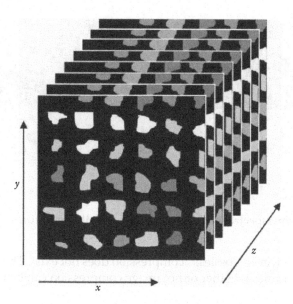

Figure 5.7 A typical image cube generated by a hyperspectral imager, with two spatial dimensions *x* and *y*, and one spectral dimension *z*.

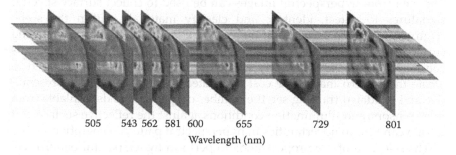

505   543 562 581 600   655   729   801
Wavelength (nm)

Figure 5.8 A hyperspectral cube obtained from imaging a grain of pollen. Multiple wavelength images are shown together with corresponding slices of spectral information. The data set is 128 × 128 × 1044 measurements (from P&P Optica, Inc.).

created over time. This visualizing method is a way to note specific spectral characteristics within the scene.

The hyperspectral cube, or hypercube, is a dynamic and powerful method to visualize data in terms of the spatial and spectral features that otherwise cannot be displayed in a single format. Using the hypercube as an analytical tool is often the first step in data analysis and exploitation. Hypercubes are automatically generated through many commercially available hyperspectral analysis software packages (Figure 5.7 and Figure 5.8).

The amount of data required to generate hyperspectral data cubes increases as the spectral and spatial dimensions decrease. Because the ability to handle these large files can be a problem, image compression is usually

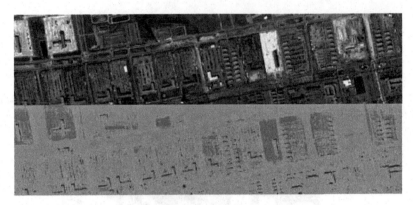

Figure 5.9   Hyperspectral pattern recognition of building features (Huertas et al., 1999).

considered. However "lossless" compression does not exist. The criticality of the data determines whether or not image compression can be used.

Pattern Recognition

The data from hyperspectral images can be used to collect surface spectral signatures to detect, identify, and classify materials within the scene (Figure 5.9). Pattern recognition methods are valuable tools for monitoring surface features, but the quality of the pattern recognition classifiers is commonly based on the quality of the training data sets that are used to create the pattern maps. The cost associated with collecting a representative and validated training set, the number of spectral bands available from most sensors, and illumination conditions against the reflective surface and terrain can lead to uncertainties in hyperspectral pattern recognition.

The selection of the proper bands to perform hyperspectral pattern recognition is extremely important. To perform the proper classification, many detection algorithms require a high level of statistical correlation. Hyperspectral data has many bands, which makes statistical significance difficult when the entire spectrum is considered. Unlike optical image pattern recognition where spatial resolution leads to better geometric shape fitting and target recognition, the primary component of spectral pattern recognition is the spectral response at a given wavelength. Spectral pattern recognition also uses spatial data for pattern recognition.

High spectral resolution can be used to trigger spatial pattern recognition algorithms. Many objects of interest consist of less than ten spatial pixels, so the sub-pixel detection algorithms are used because these algorithms increase the probability of detection for objects this small. The spectral data correlation to the target endmember provides the catalyst for automated target recognition because the object will often have symmetry or features that have been previously catalogued.

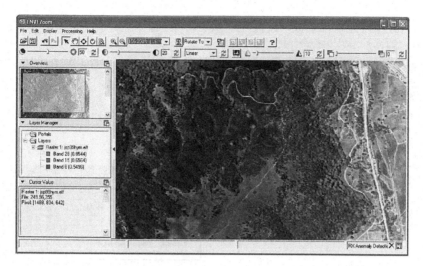

Figure 5.10 ENVI user interface and sample imagery (courtesy of ITT).

The development of spectral pattern recognition software is highly competitive among the academic and commercial sectors, and the software is generally tailored for specific applications.

Software Tools

Many software tools are available for the preparation and exploitation of hyperspectral data. Two common, commercially available analytical software tools are ENVI™ and ERDAS.

ENVI is a software tool that is used to analyze and display multispectral, hyperspectral, or radar remote sensing data. ENVI has no limits on file size or number of bands that can be analyzed during any given processing session. ENVI is used to process data such as LANDSAT, SPOT, and RADARSAT, while also accepting data from EarthWatch, ORBIMAGE, and Space Imaging. Figure 5.10 shows a user interface and sample imagery window from ENVI.

ENVI has tools to extract spectra, reference spectral libraries, and analyze high spectral resolution image datasets from many different sensors. ENVI is written entirely in Interactive Data Language (IDL), which is an array-based language that provides integrated image processing and display capabilities.

ERDAS from Leica Geosystems GIS and Mapping, LLC and the corresponding spectral analysis tool IMAGINE® contain algorithms and other industry-recognized preprocessing techniques for hyperspectral data analysis. ERDAS also processes information from many sensor systems including AVIRIS and Hyperion. ERDAS creates material mapping information from the spectral data with minimal user interaction. Figure 5.11 shows a user interface and sample imagery window from ERDAS.

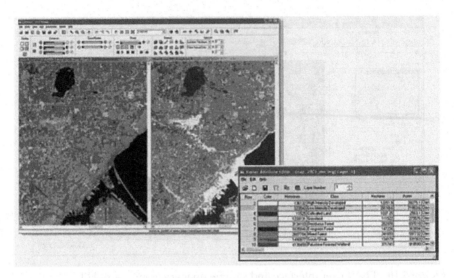

Figure 5.11 ERDAS Imagine user interface and sample imagery (courtesy of Leica Geosystems GIS and Mapping, LLC).

The U.S. government has also developed a special purpose remote sensing analytical software package called COMET for the analysis of synthetic aperture radar, multispectral and hyperspectral, overhead non-imaging, and other sources of geospatial and electromagnetic spectrum data. This software is not available for use by the general public.

Before the hyperspectral data can be processed, the data must be normalized for the effects of atmospheric transmission. The U.S. Air Force collaborated with industry to develop the MODerate spectral resolution atmospheric TRANsmittance (MODTRAN) algorithm. MODTRAN calculates atmospheric transmittance and radiance. MODTRAN incorporates most of the capabilities of LOWTRAN, which is a computer code that is widely used to calculate atmospheric transmittance and/or radiance in the infrared, visible, and near ultraviolet spectral regions. MODTRAN can calculate spherical refractive geometry, solar and lunar source functions, scattering (Rayleigh, Mie, single, and multiple), and default profiles (gases, aerosols, clouds, fogs, and rain).

The Air Force also sponsored the development of the HIgh-resolution TRANsmission (HITRAN) molecular absorption database. HITRAN is a spectral database with computational code used to predict the transmission and emission of light in the atmosphere. HITRAN currently contains over 1,700,000 spectral lines for 37 different molecules.

ATmospheric CORrection (ATCOR), which was developed for use in ERDAS, takes into consideration terrain and illumination conditions and attempts to present a more realistic response of the spectral signal as it reflects off surfaces. ATCOR significantly reduces atmospheric and illumination effects in spectral imagery.

Another atmospheric correction software package is Atmospheric CORrection Now (ACORN), which provides an atmospheric correction of spectral data from 350 to 2,500 nm. ACORN uses look-up tables calculated with the MODTRAN 4 radiative transfer code to model atmospheric gas absorption, as well as molecular and aerosol scattering effects. These modeled atmospheric characteristics are used to convert the calibrated sensor radiance measurements to apparent surface reflectance. The technique uses a fast and accurate look-up table to calculate water vapor amounts on a pixel-by-pixel basis. The user can choose to use the water vapor absorption bands at 940 nm, 1,150 nm, or both for water vapor derivations. The user can also input a visibility parameter or ask ACORN to estimate the visibility from the data. A set of sophisticated artifact suppression options are included in the ACORN software.

HATCH (High-Accuracy Atmosphere Correction for Hyperspectral data) is also used for specific atmospheric applications. Because HATCH specifically targets the atmospheric radiative transfer problems in visible and short wave infrared (SWIR) regions only, a radiative transfer algorithm is used rather than the general purpose atmospheric transmission code MODTRAN, which speeds up the data processing (Berk et al., 1989). HATCH allows the user to define the mixture of aerosols for a more accurate account of the losses associated with atmospheric transmission.

MATLAB® is a computer language used to develop algorithms, interactively analyze data, view data files, and manage projects. MATLAB solves technical computing problems faster than traditional programming languages, such as C, C++, and Fortran, and MATLAB code can be integrated with other languages and applications.

MATLAB includes development tools that implement algorithms, including the MATLAB Editor, which provides standard editing and debugging features, such as setting breakpoints and single stepping; M-Lint Code Checker, which analyzes the code and recommends changes to improve its performance and maintainability; MATLAB Profiler, which records the time spent executing each line of code; and Directory Reports, which scan all the files in a directory and report on code efficiency, file differences, file dependencies, and code coverage.

The graphics features that are required to visualize hyperspectral data are available in MATLAB. These include 2-D and 3-D plotting functions, 3-D volume visualization functions, and tools for interactively creating plots. You can customize plots by adding multiple axes, changing line colors and markers, adding annotation and legends, and drawing shapes. Figure 5.12 shows a collection of graphs constructed interactively in MATLAB by dragging data sets onto the plot window, creating new subplots, changing properties such as colors and fonts, and adding annotation.

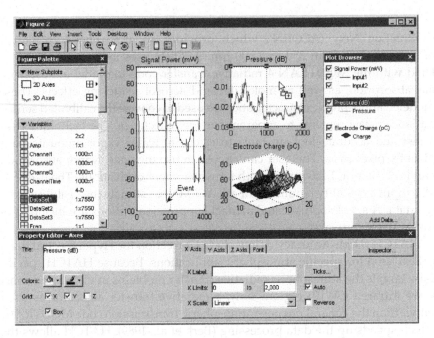

Figure 5.12 A sample window in MATLAB (courtesy of Mathworks).

References

Adams, J. B., M. O. Smith, P. E. and Johnson. 1986. Spectral mixture modeling: A new analysis of rock and soil types at the Viking Lander 1 site. *Journal of Geophysical Research*, 91(B8), 8090–8112.

Arya, S. P. 2001. *Introduction to Micrometeorology*, Academic Press, New York, 450 pp.

Berk, A., L.S. Bernstein, and D.C. Robertson. 1989. MODTRAN: A moderate resolution model for LOWTRAN 7, Final Report, *GL-TR-0122*, AGFL, Hanscom AFB, MA, 42 pp.

Clark, R. N. 1999. Chapter 1: Spectroscopy of Rocks and Minerals, and Principles of Spectroscopy, In: *Manual of Remote Sensing, Volume 3, Remote Sensing for the Earth Sciences*, A. N. Rencz, ed., John Wiley and Sons, New York, 3–58.

Clark, R. N., and T. L. Roush. 1984. Reflectance spectroscopy: Quantitative analysis techniques for remote sensing applications, *Journal of Geophysical Research*, 89, 6329–6340.

Clark, R. N., and G. A. Swayze. 1995. Mapping minerals, amorphous materials, environmental materials, vegetation, water, ice, and snow, and other materials: The USGS Tricorder Algorithm. In: Summaries of the Fifth Annual JPL Airborne Earth Science Workshop, JPL Publication 95-1, 1, 39–40.

Clark, R. N., G. A. Swayze, K. E. Livo, R. F. Kokaly, S. J. Sutley, J. B. Dalton, R. R. McDougal, and C. A. Gent. 2003. Imaging spectroscopy: Earth and planetary remote sensing with the USGS Tetracorder and expert systems: *Journal of Geophysical Research*, 108(12), 5-1–5-44.

Dwivedi, R. S., and B. R. M. Rao. 1992. The selection of the best possible TM band combination for delineating salt-affected soils. *International Journal of Remote Sensing*, 13, 2051–2058.

Huertas, A., R. Nevatia, and D. Landgrebe. 1999. Spectral Mapping Use of Hyper-spectral Data with Intensity Images for Automatic Building Modeling, in: *Proceedings of the Second International Conference on Information Fusion*, Sunnyvale, CA, July, 680–687.

Kruse, F. A., A. B. Lefkoff, J. W. Boardman, K. B. Heidebrecht, A. T. Shapiro, J. P. Barloon, and A F. H. Goetz. 1993. The spectral image processing system (SIPS)—Interactive visualization and analysis of imaging spectrometer data, *Remote Sensing of Environment*, 44, 145–163.

Price, J. C. 1994. Band selection procedure for multispectral scanners, *Applied Optics*, 33(15), 3281–3288.

6

Hyperspectral and Ultraspectral Information Extraction Approaches

Hyperspace

Most people understand the difference between a two-dimensional and a three-dimensional object. A square has two dimensions and a cube has three dimensions. Not as commonly understood is a hypercube, which is a cube with four dimensions. Just as it would be difficult to visualize a cube from a two-dimensional drawing if you had never actually seen a cube, visualizing a hypercube is also difficult.

One of the simplest ways to view higher dimensions is by slicing. If you slice a cube parallel to its sides, you would see a square, a two-dimensional figure. If you slice a hypercube, which is the cubic equivalent in four dimensions, you would see a cube.

Another way to define a hypercube is by its geometric properties. A square has a perimeter and an area. The cube, however, is defined geometrically by the intersection of six squares, also perpendicular and parallel to each other. Now instead of a perimeter, the cube has surface area and volume. The hypercube is an intersection of multiple cubes, and instead of surface area and volume, the hypercube has hypervolume, which is hard to interpret because it is defined as the side length to the fourth power. Hypervolume cannot be applied, however, because hypercubes do not exist literally.

Table 6.1 shows that when a new dimension is added, one geometric property from the previous dimension is lost and an entirely new property is gained (Huq, 2002).

Table 6.1 Geometric Properties for Different Dimensions

Shape	No. of Dimensions	Geometric Property
Square	1 dimension	Perimeter
	2 dimensions	Area
Cube	2 dimensions	Surface area
	3 dimensions	Volume
Hypercube	3 dimensions	Spatial volume
	4 dimensions	Hypervolume

Hyperspace is a topological space with a fourth dimension. The elements of hyperspace are subsets of another topological space.

With respect to hyperspectral remote sensing, each hyperspectral band corresponds to a hyperspace dimension. Data from a 64-band sensor would be described by a 64-dimension hyperspace.

The Importance of Endmembers

Chapter 5 discussed how the spectral signatures for endmembers are collected and stored. Detection algorithms can analyze a complex mixture of signatures and identify the individual endmembers that make up the signature because all endmembers, or reference spectra, of the same object will have the same distinct spectral signature. These reference spectra are then stored in a spectral library.

For example, when hyperspectral data is collected over a geographical area, the spectral signatures for the vegetation, soil, water, and rocks within the area will be contained in the collected data. Once the detection algorithms unmix all the signatures, the individual reference spectra will identify the objects on the ground.

Spectral Libraries

The FBI maintains a database for fingerprints, and anyone who has been fingerprinted for any reason might have their fingerprints in his or her file. When a specific unknown person needs to be identified, that fingerprint is sent to the FBI database and analyzed to find a possible match. If that person's fingerprint is in the database, then the unknown person is identified.

A spectral library is similar to the fingerprint database except instead of containing human fingerprints, the spectral library contains spectral signatures, the "fingerprints" that are unique to materials on the Earth's surface. A specific healthy tree variety, for example, will have the same spectral signature as every other healthy tree of that same variety. Over the years, researchers have been collecting spectral signatures of known objects and cataloguing them in spectral databases or libraries.

After the hyperspectral data from a given scene have been analyzed and the spectral signatures of the objects have been identified, these signatures can be compared with those in a spectral library and the objects can be identified. Image processing software packages include vast spectral libraries, but because the spectral signatures for every object in the world have not been collected, users often need to add new signatures to facilitate the spectral identification process.

Delineation vs. Identification

Reference spectra can be derived from a hyperspectral image and used to identify the features of interest within the study area. Once the features of interest have been identified on a large scale, the data can be further analyzed to determine the specific location or boundaries of each object. This process is called "delineation."

Pixel Unmixing (Abundances)

The spectral data collected from a geographic area is a linear combination of the spectra of all the materials that appear in the image. Performing a least-squares fit will determine the weighting coefficients for the spectrum for each individual material, which gives the best fit to the original spectrum. The weighting coefficients are considered to be equal to the abundances of the respective materials (Ray, 1994).

Linear and Nonlinear Mixing

A lot of remote sensing analysis has been based on the concept of the Earth as spots covered by differently colored paint. When the spots of paint get too small, they appear to blend together to form a new color, which is a simple mixture of the old colors. For example, if an area is covered by 50% small red spots and 50% small green spots, from far away, the surface appears as yellow. Different proportions of red and green dots will produce different colors. If you know that the surface is covered by red and green dots, the proportions of the colors can be calculated based on the color that you see. Any light reaching the observer, however, has only illuminated one of the colored dots. That is linear mixing.

Nonlinear mixing occurs when light hits more than one of the colored dots and the spectra of the materials being observed are twisted into different spectra that do not resemble any of the targets. For nonlinear mixing, instead of flat dots, imagine a surface with a lot of small, colored bumps that stick out varying distances from the surface. Light would bounce from one colored bump to another and then to the observer. Because some of the light coming from a green bump bounced off a red bump first, this light would have characteristics of both the red and green bumps. However, some of the light comes directly from the green bump that only bounced from the green bump. If you could see this individual green bump, it would not look as green as it should. When the light from all of the bumps reaches the observer, the light looks different than when the bumps were simple spots, even through the proportion of the area covered by each color is unchanged, assuming that there are no shadows.

A second way for nonlinear mixing to happen is if light passes through one material and then reflects off another. Imagine a piece of translucent plastic with half of the area covered by randomly placed translucent green spots placed on top of a red surface. Now light can pass through a green spot on the plastic and then reflect off the red below before returning to the observer. Once again, the interaction of the light with multiple spots along its

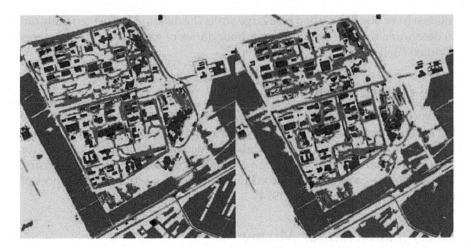

Figure 6.1 Change detection from Quickbird data between 2002 (left) and 2003 (right) (Niemeyer and Nussbaum, 2005).

path changes the character of the light coming from each spot. Once again, the color looks different than it does in the linear case (Ray, 1994).

Change Detection

One of the powerful analytical tools that can be performed with the information contained in hyperspectral imagery is change detection. Many times the changes are nearly undetectable either because of the slightness of the change or more importantly, because the change is outside the visible spectrum of the human eye. By using two or more spectral images or data from the same geo-location but collected at different times, the spectral change within the scene can be assessed (Figure 6.1).

Remote sensing considerations for spectral change detection include the time and look angle (the solid angle in which an instrument operates effectively) from which the data was collected. Although shading does not change the spectral response, shading does have an impact on the magnitude of the signature, which can impact the change detection algorithms. The offset angle of the remote sensor data collection can impact the pixel geometry. Near-nadir collections are best suited for change detection analysis so that shading is minimal.

Algorithms and comparison routines can refine the exploitation by showing only the changed or affected areas. Dithering between two images is also commonly used to visually determine changes between different images in the sequence.

Spectral Maps

After hyperspectral data are converted into hyperspectral information, the result is often displayed in a spectral map. Many types of maps can be

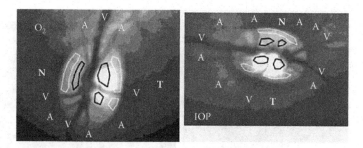

Figure 6.2 Single-band images (570 nm) from hyperspectral data used for mapping the optical nerve head and vessels for oxygen breathing (left) and interocular pressure (right) (Khoobehi et al., 2004).

Figure 6.3 AVIRIS false-color composite, 27 June 2001, near Mammoth Mountain, Sierra Nevadas, with vegetation mapping in green and orange (Maurer, 2002).

generated from spectral data including visible, gray scale, false infrared, and other false color schemes to assist with data exploitation. Hyperspectral mapping applications include agriculture and forestry, mineral and gas exploration, pollution control, and national defense.

Although many remote hyperspectral sensing applications are macroscopic with respect to surface features of the Earth, applications in the biomedical fields require positional information on a microscopic level, such as images obtained within the human eye for analysis and treatment, as shown in Figure 6.2.

More traditional spectral maps include forestry (Figure 6.3) and mineral (Figure 6.4) mapping. The quality of the data provided by hyperspectral sensors significantly increases the resolution of the boundaries and provides the ability to perform spectral matching and change detection. It also significantly reduces the amount of time required to collect data from ground-based collection systems.

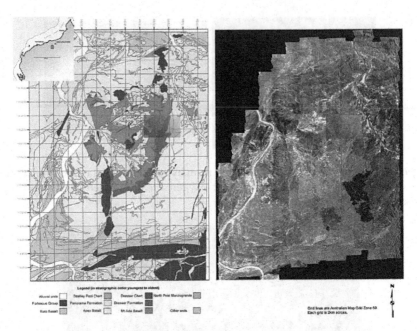

Figure 6.4 Simplified geological map (left) of the North Pole Dome (modified after Van Kranendonk, 2000). Visible wavelengths true color (red: 0.6069 mm, green: 0.5615 mm, blue: 0.4693 mm) image of the hyperspectral dataset (left) (Brown et al., 2005).

References

Brown, A. J., M. R. Walter, and T. J. Cudahy. 2005. Hyperspectral imaging spectroscopy of a Mars analogue environment at the North Pole Dome, Pilbara Craton, Western Australia, *Australian Journal of Earth Sciences*, 52, 353–364.

Huq, R. 2002. *Temporal Annex*. University of Maryland at College Park, http://temporal_science.tripod.com.

Khoobehi, B., J. M. Beach, and H. Kawano. 2004. Hyperspectral imaging for measurement of oxygen saturation in the optic nerve head, *Investigative Ophthalmology and Visual Science*, 45, 1464–1472.

Maurer, J. 2002. Mapping ponderosa and Douglas-fir using AVIRIS: An exploratory analysis in the Sierra Nevadas, Department of Geology, University of Colorado at Boulder, May, p. 1.

Niemeyer, I, S. Nussbaum, and M. J. Canty. 2005. Automation of change detection procedures for nuclear safeguards-related monitoring purposes, Geoscience and Remote Sensing Symposium Proceedings, IEEE International, 3(25–29) 2133–2136.

Ray, T. W. 1994. *Vegetation Remote Sensing FAQs*, Div. of Geological and Planetary Sciences, California Institute of Technology, www.yale.edu/ceo/Documentation/rsvegfaq.html.

Van Kranendonk, M. J. 2000. Geology of the North Shaw 1:100 000 Sheet. Geological Survey of Western Australia, Perth.

7

Agricultural Applications

Farmers can order spectral imagery of their fields to determine the status of their land and whatever is growing on it. For example, spectral imagery can indicate the amount of fertilization required in specific locations that are designated with GPS coordinates. Agricultural machinery on the market today has the capability to load this information into computers built into the machinery and automatically adjust the amount of fertilizer deposited based on the information contained in the spectral imagery.

The type of vegetation can also be determined from spectral remote sensing. Figure 7.1 (Clark et al., 1995) shows the types of plants growing in a field. Because stressed vegetation looks different from healthy vegetation, mapped remote sensing information can be an indication of plant disease or drought. LANDSAT Thematic Mapper™ has been used to determine stress in trees caused by salinity in the soil (Dwivedi and Rao, 1992). New, more powerful hyperspectral sensors are now being flown over citrus groves in Florida and other parts of the Southeast in search of citrus canker, the new plague of the

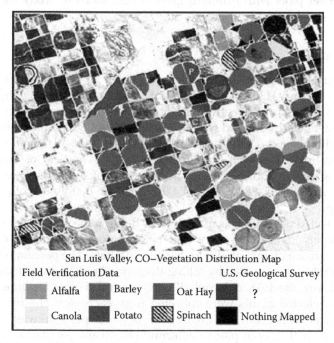

Figure 7.1 Vegetation distribution map (Clark et al., 1995).

market that is responsible for the trees being quarantined and destroyed. The significance of this contamination was highlighted by reports that indicate that since 2003, Florida was responsible for producing over 86% of the orange juice consumed in the United States (Brown and Brown, 2001).

Case Study 1: Detecting Crop Growth Parameters

Case Study 1 was conducted by Patel et al. (2001) to determine if high-spectral-resolution remote sensing data could be used to detect the variation in crop growth parameters such as LAI, chlorophyll content, and biomass. The researchers looked at the position of the red edge for wheat crops growing under different conditions. Then they studied the relationship between the crop growth parameters and the inflection wavelength, which was computed from the red-edge parameters.

Study Area

Patel et al. (2001) selected a wheat-growing area of Dholka Taluka (administrative block) of the Ahmedabad district, Gujarat state that is irrigated by the Fatehwadi canal. The wheat crop was sowed in the middle of October and harvested in the middle of March. An area of approximately 10 × 2 km was selected for the Airborne Imaging Spectrometer (AIS) flight. The researchers picked wheat plots with different growth conditions to study the growth parameters of the crop.

Data Used

High-spectral-resolution data were acquired on 24 and 26 February 1997 between 1100 and 1130 hrs from a flight height of 3 km using the AIS. At the same time that the spectral data of selected plots were collected in the airplane, data were collected from the ground level with a portable Ground Truth Radiometer (GTR) developed by the Space Applications Center of the Indian Space Research Organization (ISRO). The field of view of the radiometer was 15 and the central wavelengths of 10 bands were 490, 565, 660, 670, 710, 745, 785, 880, 960, and 1,025 nm. The bandwidth of each band was 10 nm.

The reflected radiance was measured by holding the radiometer vertically approximately 1.5 m above a plot, and measurement of irradiance was made using a reference plate coated with barium sulphate. The percent reflectance in each band was calculated by taking the ratio of radiance to irradiance.

To measure the LAI, chlorophyll content, and biomass, samples of ten randomly selected plants were collected from each plot. The fresh weight of these plants was taken and the number of plants per 1 m^2 was counted at three places within each plot to determine the plant population. Leaf area was measured using a leaf area meter (LI COR model LI 300). Leaf area index

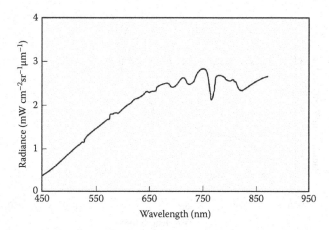

Figure 7.2   Spectral radiance of soil. (Patel et al., 2001.)

(LAI) and biomass were calculated using measurements of sample plants and plant populations.

Data Analysis

The data of Visalpur village were extracted and converted into a sub image of 143 bands. A false color composite (FCC) was prepared using one each in green, red, and near-infrared bands and used to identify selected plots. Signatures of selected plots were generated in all bands, and mean values of digital counts were obtained. Radiometric calibration constants of gain and offset for each band, provided along with the raw data, were used to convert digital counts into radiance.

Wheat plots at different growth stages, along with bare soil (fallow plot), were selected to study spectral characteristics. Figure 7.2 and Figure 7.3 show the spectral radiance response of soil and wheat, respectively. Wheat shows low radiance in the red region and higher radiance in the near-infrared region compared to soil. The absorption dip observed at 760 nm is mainly due to absorption by atmospheric oxygen (Bach et al., 1995).

The spectral reflectance of the soil (fallow plot) measured by radiometer was assumed to be the same at the altitude of the AIS. Considering soil reflectance as reference, irradiance was calculated and further used to calculate percent reflectance from selected plots. Figure 7.3 shows absorption by the photosynthetic active biomass of the crop canopy in the red region near 670 nm and high reflectance in the near-infrared region above 780 nm.

The position of the red edge is determined by the inflection wavelength, which is defined as the wavelength at which the rate of increase of reflectance is the maximum. The position and shape characteristics of the red edge in the visible and near infrared are good indicators of plant parameters. Miller et al. (1991) evaluated an inverted Gaussian model for the vegetation red-edge reflectance.

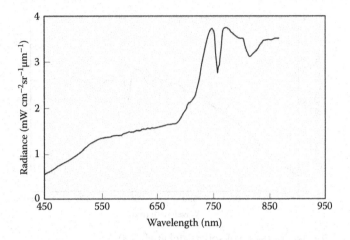

Figure 7.3 Spectral radiance of wheat crop. (Patel et al., 2001.)

Fitting of inverse Gaussian function to the spectral data in this region leads to four parameters that represent the red-edge characteristics (Bach and Mauser, 1991). Spectral reflectance curves for vegetation exhibit a consistent shape in the red-edge region characterized by relatively broad flat minima in the 670 nm region, followed by a sharp increase in reflectance beginning at about 685 nm and an asymptotic reflectance plateau reached at wavelengths beyond 780 nm. The inverted Gaussian model, which represents the red edge by the reflectance, is:

$$R(\lambda) = R_s - (R_s - R_0)\exp\left(\frac{-(\lambda_0 - \lambda)^2}{2\sigma^2}\right)$$

where R_s is the maximum or shoulder spectral reflectance, R_0 is the minimum spectral reflectance corresponding to the chlorophyll absorption well, λ_0 is the central minimum or peak absorption wavelength, λ is the wavelength in the red and red-infrared region, and σ is the Gaussian function deviation parameters. A fifth parameter is λ_p, the wavelength of inflection of the red reflectance edge slope, defined by the wavelength of the maximum in the first derivative of the Gaussian function (Miller et al., 1990):

$$\lambda_p = \lambda_0 + \sigma$$

The parameter λ_p provides another measure of the position of the vegetation red reflectance edge.

The standard numerical procedure was used to produce a best fit to the reflectance data according to the least-square criterion. The details of the procedure are described by Bonham-Carter (1988). For inverted Gaussian model

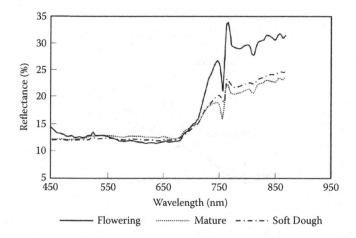

Figure 7.4   Spectral reflectance of wheat at different growth stages. (Patel et al., 2001.)

fitting, reflectance data from 650 to 780 nm were used, except data from 750 to 770 nm wavelengths due to absorption by atmospheric oxygen at 760 nm. This model was used for 17 plots of wheat with different growth conditions.

Results and Discussion

The red edge is a phenomenon caused by the combination of the chlorophyll absorption in the red region and the scattering by the plant cells in the near-infrared region of the electromagnetic spectrum. The wavelength position of the red edge is a parameter that is sensitive to plant development (Guyot et al., 1992).

The spectral reflectance curves of three wheat plots at different growth stages: flowering, soft dough, and maturity stages, are shown in Figure 7.4. The difference in crop growth is reflected in spectral responses around the chlorophyll absorption in the red and near-infrared regions. Separation between the spectral response of the wheat plots at different growth stages increased in the region of high reflectance from 750 to 880 nm. In the near-infrared region, maximum reflectance was recorded from plot A, which was at the flowering stage, and more chlorophyll content and minimum reflectance were recorded from plot B, which was at the maturity stage. This trend was reversed at the shorter wavelengths, where the wheat plot at flowering stage exhibited enhanced absorption. Plot C shows the intermediate values. This data indicates the change of reflectance values with crop growth.

The position of the red edge is determined by inflection wavelength, which is the wavelength at which the rate of increase of reflectance is the maximum. Through Gaussian curve fitting to the reflectance data, the reflectance wavelength can provide an effective quantitative representation of the

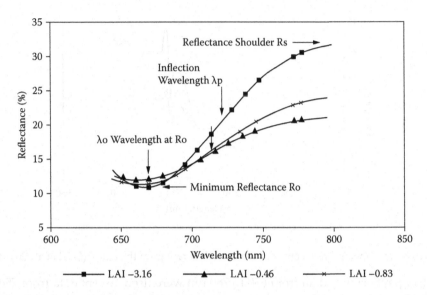

Figure 7.5 Inverted Gaussian fit of wheat spectra. (Patel et al., 2001.)

shape and position of vegetation red-edge reflectance in terms of physical significance (Miller et al., 1991).

The inflection wavelength of selected wheat plots was calculated using an inverted Gaussian fit. Figure 7.5 shows the graph for three wheat plots at different LAIs. Modeling of the red edge shows longest inflection wavelength of plot A, with an LAI of 3.16. The reduction in LAI and chlorophyll in the senescence phase is reflected in the graph line of plots C and B, with LAIs of 0.83 and 0.46, respectively. The shift of the inflection wavelength is marked in Figure 7.5. The chlorophyll content of plot W1024 was higher compared to the other two plots, which indicates that the position of the inflection wavelength shifted toward a longer wavelength as the LAI increased.

The researchers made an attempt to derive a relationship between inflection wavelength and crop growth parameters. The scatter plot of inflection wavelength with LAI and chlorophyll content is shown in Figure 7.6 and Figure 7.7, respectively. Regression analysis was carried out and a linear relationship was observed between the inflection wavelength and the LAI and chlorophyll content. The wheat crop was in the post-heading stage at the time of the AIS test flight in the month of February but, along with green leaves, dry leaves, and other components, also contributed to the spectral response, resulting in low values of correlation coefficients.

Conclusion

This experiment indicates that at least some crop growth parameters can be determined using AIS data. In this case study, the inflection wavelength of

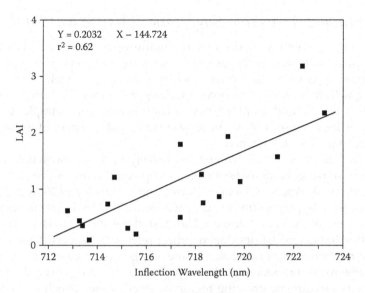

Figure 7.6   Relationship between inflection wavelength of wheat spectra and LAI. (Patel et al., 2001.)

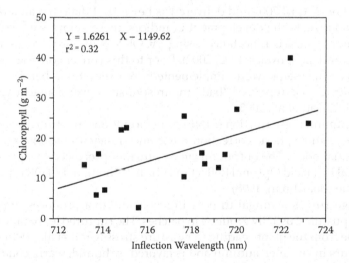

Figure 7.7   Relationship between inflection wavelength of wheat spectra and chlorophyll content. (Patel et al., 2001.)

the spectra of wheat shifted from 713 to 723 nm at different growth stages. Patel et al. determined this shift by extracting the inflection wavelength of the red edge of the reflectance spectra. A linear relation of the inflection wavelength with the LAI and chlorophyll content of wheat indicates that high-spectral-resolution data can be used to assess crop growth conditions and identify stressed crops.

Case Study 2: Detecting Sugarcane "Orange Rust" Disease

Disease management is important in maintaining the competitive advantage of the sugar industry. Pathogens can cause serious damage to sugarcane (*Saccharum* spp.) crops that often lead to reduced crop yield and quality. Dealing with this problem involves a variety of curative measures, in which disease detection and mapping play a central role. For example, to apply chemicals for disease control, the location and spatial extent of the affected crops must be first determined.

Thenkabail et al. (2002) worked extensively with narrow-band spectral indices to make general assessments of crop conditions. In this case study conducted by A. Apan, A. Held, S. Phinn, and J. Markley in 2004, the potential of satellite hyperspectral imagery was examined to detect the incidence of sugarcane "orange rust" disease. They tested the utility of existing spectral vegetation indices (SVIs), developed indices relevant to disease detection, and studied relationships between sugarcane orange rust disease and changes to the biochemical component of the crop. Their study area covered a portion of Mackay's sugarcane growing region in Queensland, which is the largest sugar-producing area in Australia.

The researchers used an image from the Hyperion sensor on EO-1 acquired on 2 April 2002 and delivered as Level 1B_1 data in scaled radiance units. To facilitate the development of indices, these values were converted to apparent surface reflectance using ACORN 4.10 software (Analytical Imaging and Geophysics, LLC, 2002). Prior to this conversion, the following pre-processing steps were implemented: re-calibration, band selection, de-streaking, and repair of "bad" (nonresponsive) pixel values (Apan and Held, 2002; Datt et al., 2003).

A minimum noise fraction (MNF) transformation smoothing was applied to the post-atmospheric correction reflectance image to minimize uncorrelated spatial noise. The output image was further processed by applying the Empirical Flat Field Optimal Reflectance Transformation (EFFORT) polishing technique (Boardman, 1998).

Orange rust is a fungal disease in sugarcane that produces orange leaf lesions (pustules) usually grouped in patches. The ruptured leaves allow water to escape from the plant, leading to moisture stress (Croft et al., 2000). Orange rust occurs in summer/autumn and is favored by humid, warm conditions.

In this case study, the infected fields were rated at the canopy level as 4 based on our 1 to 5 scale (1 having lowest severity to 5 having highest severity). Referenced at the time of Hyperion overpass on 2 April 2002, the information on the location and severity rating of orange rust was sourced from the field offices of the Mackay Sugar cooperative.

Diagnostic symptoms of orange rust in image datasets could be related to changes in leaf pigments, internal leaf structure, and moisture content. Therefore, SVIs focusing on one or more attributes associated with these symptoms were selected. Although the majority of indices were sourced from the literature, five indices were formulated during this case study based on the examination of detailed spectral reflectance plots.

Statistical Analysis

Polygons were digitized around the sugarcane blocks affected with the orange rust disease, as well as several blocks not affected by the disease, to produce 142 and 159 sample pixels, respectively. The nondisease blocks contained the same variety (Q124) and age group of sugarcane as the diseased blocks. A discriminant function analysis was used to generate discriminant functions based on linear combinations of Hyperion band indices that provided optimum discrimination between rust-affected and non-rust-affected areas (SPSS, 2001). The accuracy of the model was evaulated by classifying a "hold-out sample" (i.e., those pixels not included in model generation) corresponding to 30% of the total sample pixels.

Results and Discussion

Reflectance spectra of Hyperion "raw" bands showed that sample areas with the sugarcane orange rust disease exhibited different spectral reflectance signatures and could be discriminated from nondiseased areas at certain wavelengths (Figure 7.8). The highest separability was located in the near-infrared (NIR) region (between 750 to 880 nm and 1,070 nm). This level of separability was followed by selected ranges in the short wave infrared (SWIR; 1,660 nm and 220 nm), green (550 nm), and red (680 nm) regions. Disease-affected areas had relatively lower reflectance values than unaffected sites in the green and NIR regions. However, the reverse was true for the red and the SWIR domains where areas with orange rust had higher reflectance values than areas of sugarcane with no rust.

The results of the discriminant function analysis indicate the following:

- The 1,600 nm (SWIR) band, if combined by ratioing with either NIR band (800 nm) or green band (550 nm), produced the best results (i.e., the largest correlation with the discriminant function and the highest classification accuracy) among the indices. This was the case for the four highest ranked indices (DWSI-1, DSWI-2, DSWI-5, and MSI).
- The indices that only incorporated selected bands in the very near infrared (VNIR) (e.g., Ave (750–850, SIPI, DSWI-4, ND800/600, OSAVI, TCARI, PSSRa) performed moderately.
- The indices developed from the reflectance red edge (69–720 nm) (e.g., REIP-Lagr and REIP-poly) were relatively poor in discriminating diseased from nondiseased sugarcane crops. They produced very small correlations with the discriminant function, and their classification accuracies were among the lowest.

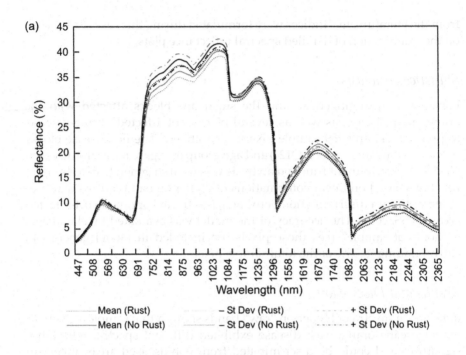

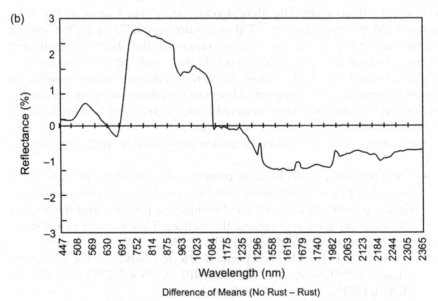

Figure 7.8 Reflectance spectra of Hyperion sample pixels containing sugarcane orange rust disease and without orange rust disease: (a) mean and standard deviations and (b) difference of means. (Apan et al., 2004.)

- The output discriminant function, a linear combination of DSWI-2, SR695/420 and NDWI-Hyp, attained a classification accuracy of 96.9% for the hold-out sample pixels.

The results of this study showed that if moisture-sensitive bands in the SWIR region were incorporated, the spectral discrimination of sugarcane that had a moderate to high severity of orange rust disease could be significantly increased. The loss of moisture due to lesions or ruptured leaves played an important factor in the disease detection (Croft et al., 2000). The high levels of discrimination provided by the selected disease–water stress indices developed from this case study reinforced this point. However, crops that were water-stressed, but not necessarily afflicted with orange rust disease, could potentially be differentiated from these indices. Such conditions would complicate orange rust detection.

Conclusion

Several narrow-band indices derived from Hyperion image data were able to discriminate sugarcane crops that were severely affected by orange rust disease from nondiseased areas in the Mackay region of Australia. These indices used spectral bands that are known to be sensitive to changes in leaf pigments, internal leaf structure, and moisture content. The discriminate function analysis allowed the researchers to rank each index based on its ability to differentiate rust-affected vs. non rust-affected pixels. Although the VNIR-based indices offer significant separability, the incorporation of a 1,660-nm SWIR band that led to the formulation of the disease–water stress indices provided the maximum discrimination.

References

Apan, A., and A. Held. 2002. *In-house Workshop on Hyperion Data Processing: Echoing the Sugarcane Project Experience,* CSIRO Land and Water, Black Mountain Laboratories, Canberra.

Apan, A., A. Held, S. Phinn, and J. Markley. 2004. Detecting sugarcane "orange rust" disease using EO-1 Hyperion hyperspectral imagery, *International Journal of Remote Sensing,* 25(2), 489–498.

Bach, H., and W. Mauser. 1991. Extraction of agricultural parameters from imaging spectrometry data and simulated MERIS data through red edge analyses. European Imaging Spectroscopy Aircraft Campaign. ESA SP-360 (Noordwijk: ESA).

Bach, H., A. Demircan, and W. Mauser. 1995. The use of AVIRIS data for the determination of agricultural plant development and water content. *Proceedings of MAC Europe '91—Final Results,* ESA WPP-88 (Paris: ESA), Lenggries, Germany.

Boardman, J. W. 1998. Post-ATREM polishing of AVIRIS apparent reflectance data using EFFORT: A lesson in accuracy versus precision. In: *Summaries of the Seventh JPL Airborne Earth Science Workshop,* JPL, Pasadena, CA, 1, 53.

Bonham-Carter, G. F. 1988. The numerical procedures and computer program for fitting an inverted Gaussian model to vegetation reflectance data, *Geoscience*, 14, 339–356.

Brown, C. A., and M. G. Brown. 2001. Florida Citrus Outlook for 2003–2004 Season. Presented to the Florida Citrus Commission, Lakeland, Florida, March 20, 25 pp.

Clark, R. N., T. V. V. King, C. Ager, and G. A. Swayze. 1995. Initial vegetation species and senescence/stress mapping in the San Luis Valley, Colorado using imaging spectrometer data. *Proceedings: Summitville Forum '95*, H. H. Posey, J. A. Pendelton, and D. Van Zyl, Eds., Colorado Geological Survey Special Publication 38, 64–69.

Croft, B., R. Magarey, and P. Whitle. 2000. Disease management. In: *Manual of Cane-growing*, edited by M. Hogarth and P. Allsopp, Bureau of Sugar Experiment Stations, Brisbane, 263–289.

Datt, B., T. R. McVicar, T. G. Van Niel, D. L. B. Jupp, and J. S. Pearlman. 2003. Pre-processing EO-1 Hyperion hyperspectral data to support the application of agricultural indexes, *IEEE Transactions on Geoscience and Remote Sensing*, 41, 1246–1259.

Dwivedi, R. S., and B. R. M. Rao. 1992. The selection of the best possible TM band combination for delineating salt-affected soils, *International Journal of Remote Sensing*, 13, 2051–2058.

Guyot, G., F. Baret, and S. Jacuemoud. 1992. Imaging spectroscopy for vegetation studies. In: *Fundamentals and Prospective Applications*, edited by F. Toselli and J. Bodechtel, Kluwer Academic, 145–165.

Kanemasu, E. T., C. L. Niblett, H. Manges, D. Lenhert, and M. A. Newman. 1974. Wheat: Its growth and disease severity as deduced from ERTS-1, *Remote Sensing of Environment*, 3, 255–260.

Miller, J. R., E. W. Hare, and J. Wu. 1990. Quantitative characterization of vegetation red edge reflectance: An inverted-Gaussian reflectance model, *International Journal of Remote Sensing*, 2, 1755–1773.

Miller, J. R., J. Wu, M. G. Boyer, M. Belanger, and E. W. Hare. 1991. Seasonal patterns in leaf reflectance red edge characteristics, *International Journal of Remote Sensing*, 12, 1509–1523.

Patel, N. K., C. Patnaik, S. Dutta, A. M. Shekh, and A. J. Dave. 2001. Study of crop growth parameters using airborne imaging spectrometer data, *International Journal of Remote Sensing*, 22(12), 2401–2411.

SPSS. 2001. *SPSS for Windows* (Release 11) (Chicago: SPSS).

Thenkabail, P., R. Smith, and E. Depauw. 2002. Evaluation of narrowband and broad-band vegetation indices for determining optimal hyperspectral wavebands for agricultural crop characterization, *Photogrammetric Engineering and Remote Sensing*, 68, 607–621.

8

Environmental Applications

Hyperspectral remote sensing can be used to study the state of our environment and track changes that occur over time. This technology has been particularly successful in monitoring bodies of water of all sizes from ponds to oceans and brooks to rivers. This chapter includes a description of the work conducted by researchers for two studies, one dealing with classifying the quality of the water in a lake and the other with mapping the location of submerged aquatic vegetation.

Case Study 1: Classifying Lake Water Quality

The main advantage of using remote sensing instead of the traditional lake monitoring method based on water sample collection is its good spatial and temporal coverage. Monitoring can be carried out several times per year, and lakes too small or inaccessible to be included in the traditional sampling can be also monitored.

Introduction

After studying lakes, rivers, and coastal areas in Finland in 2002, Koponen et al. concluded that lake water quality could be classified with airborne imaging spectrometers. At the time of their study, the general water quality was periodically assessed by the Finnish Environment Administration. To classify the water quality, water samples were collected every four years from stations at selected locations. The samples were then analyzed in a laboratory.

Lake classification from samples collected during 1997 included data from 5,000 sampling stations on lakes that represent 79% of the total lake surface area of Finland (including all lakes larger than 1 km²). However, even though the collected data set is representative, its usability was limited, especially by the spatial variation of water quality in lakes.

Other researchers have discovered that by using remote sensing techniques, some of the important variables used in the operational classification of lakes can be measured. These optically active variables include chlorophyll a, total suspended solids, turbidity, and Secchi depth (see Dekker, 1993; Gitelson et al., 1993; Kallio et al., 2001). Aquatic humus is also an optically active substance sometimes used in lake classifications, but its estimation by remote sensing techniques in lakes has been proven difficult (Dekker, 1993; Kallio et al., 2001).

In the 2002 Koponen et al. study, the researchers classified the water quality using the parameters Secchi depth, turbidity, and chl-*a*. They obtained the class limits from two operational classification standards and discovered that using a combination of them was the most suitable when remote sensing data is used. Because the classification was possible even without concurrent ground truth data, they discovered that operational classification with remote sensing data is possible. Their classification accuracy ranged from 76% to 90%.

The Koponen et al. scientists also investigated the feasibility of using remote sensing data for operational lake water classification using regression algorithms. In addition, the feasibility of the Medium Resolution Imaging Spectrometer (MERIS) instrument onboard the Envisat satellite for water quality classification was studied by reconstructing the MERIS channels from airborne spectrometer data. They found that the channel configuration of the Envisat MERIS instrument also appears to be suitable for the classification of turbid lakes, such as Finnish lakes.

Instruments and Data

Koponen et al. measured eleven lakes in southern Finland during the four campaigns. The measurements were taken for eight days. The lake selected for the measurement campaigns had varying water quality characteristics. The trophic status varied from oligotrophic to eutrophic, and two of the lakes were humic. For detailed information on the lakes, see Kallio et al. (2001).

The main remote sensing instrument used during the campaigns was the AISA (Makisara et al., 1993). The main measurement characteristics of AISA are presented in Table 8.1.

AISA has a total number of 286 channels. However, the instrument is not able to store data from all channels when the measurement mode suitable for airborne remote sensing is used (the amount of data generated exceeds the capabilities of the data recorder). Instead, data from a smaller number of preselected channels are stored.

Table 8.1 Measurement Characteristics of AISA Airborne Spectrometer

Type	Pushbroom CCD-matrix sensor
Number of channels	286
Channel wavelength range	450–900 nm
Channel bandwidth	1.6–9.7 nm (sum of one to six channels)
Number of pixels (across track)	384
Field of view	21°
Pixel size from 1000-m altitude	1 m

Koponen S., J. Pulliainen, K. Kallio, M. Hallikainen. 2002. *Remote Sensing of Environment*, 79(1), 51–59. With permission.

After acquisition, the AISA images were radiometrically and geometrically corrected and resampled to a pixel size of 2×2 m by the Finnish Forest Research Institute. Additional data preprocessing consisted of deriving the average radiance of each AISA channel in a 100×100 m square around each ground truth sampling point. If the ground truth point was not at the center of the measurement swath or if it was close to shore or in a cloudy area, the square was moved to the closest suitable location at the center of the swath. Averaging reduced the variability of the signal due to the stripes caused by the CCD cell and the sun glitter caused by the rough water surface.

The ground truth measurements included water sampling for laboratory analysis (e.g., chlorophyll a, turbidity, total suspended solids, aquatic humus), on-site measurements (e.g., Secchi depth, upwelling and downwelling irradiance with an underwater spectrometer), and weather observations (e.g., wind speed and direction, cloudiness). The sum of chlorophyll a and phaeophytin a (denoted here with chl-a) was determined with the spectrophotometer after extraction with hot ethanol (ISO 10260) and turbidity by nephelometric method (based on the measurement of light, 860 nm) scattered within a 90-degree angle from the beam directed at the water sample (ISO 7027). In TP determination, the water sample was digested by potassium peroxodisulphate before analysis with ammonium molybdate (Murphy and Riley, 1962). The total number of points with near-simultaneous AISA and ground truth data was 127. Due to partial cloud cover and other problems, the number of usable data points was 122.

Methods

Retrieval Algorithms

The retrieval of water quality variables with remote sensing instruments is based on analyzing the spectral features of solar radiation reflected from the water body. The substances found in natural waters (phytoplankton, suspended inorganic material, and dissolved organic matter) scatter and absorb the incoming solar radiation. These processes, defined as the Inherent Optical Properties (IOP) by Preisendorfer (1976), are wavelength dependent and therefore influence the shape and the magnitude of the spectra reflected from water. This can be seen in Figure 8.1 where the spectra measured (by AISA) at five ground truth data points are presented. By comparing the spectra data with the water quality variables, the following features can be observed.

The peak at about 400 nm grows as the concentration of chl-a increases. This has been linked to scattering and absorption by phytoplankton (Morel and Prieur, 1977), and to chl-a fluorescence, which has a maximum at 683 nm (Smith and Baker, 1978). The shift to longer wavelengths as the concentration of chl-a increases was observed by Gitelson (1992). Just before, the peak phytoplankton has an absorption region at about 660–670 nm, although it is not as clear as the peak at 700 nm in Figure 8.1.

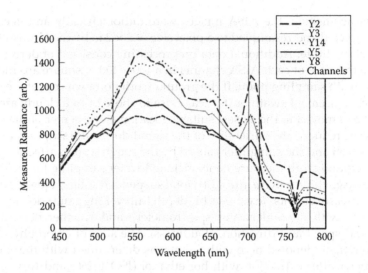

Figure 8.1 Sample spectra measured by AISA. Y2–Y14 are ground truth data points. The channels used in the retrieval algorithms are shown as vertical lines. (Koponen S., J. Pulliainen, K. Kallio, M. Hallikainen. 2002. *Remote Sensing of Environment*, 79(1), 51–59. With permission.)

Due to scattering from suspended matter, the detected radiance increases with the turbidity value in all parts of the spectrum in Figure 8.1. Since absorption by optically active substances also influences the radiance level, it must be accounted for. One way to do this is to use wavelengths where the absorption by optically active substances (e.g., chl-*a* and colored dissolved organic matter) is minimal. One such region is near 710 nm (Dekker, 1993). The data show that at that wavelength, the turbidity values follow the radiance values well by decreasing systematically with decreasing radiance.

The use of channel ratios for a relationship between remote sensing measurements and ground truth data is very common. The advantage of using ratios over absolute values of radiance (or reflectance) is that they correct some of the effects of measurement geometry and atmosphere. For example, Dekker, Malthus, and Seyhan (1991) showed that channel ratios yield high correlation coefficients for several water quality parameters. Dekker (1993) and Gitelson et al. (1993) concluded that for the retrieval of chl-*a* concentration, a ratio of channels centered at about 675 and 705 nm is useful in several lake types (oligotrophic to hypertrophic). In addition, the previous studies on partly that same data set as used here (by Kallio et al., 2001; Koponen et al., 2001; Pulliainen et al., 2001) have showed that simple channel ratio and channel difference algorithms give high coefficients of determination for the water quality variables included here.

In this analysis, the best retrieval algorithm for each variable was found empirically by deriving a regression model for all possible channels and channel ratio and channel difference combinations and selecting the one with the highest R^2.

Satellite remote sensing instruments can cover much larger areas than airborne sensors. Perhaps the most interesting satellite instrument is the MERIS on board the Envisat satellite. MERIS has several channels suitable for the estimation of water quality variable, and it has a fairly good spatial resolution of 300 m (Rast, Bezy, and Bruzzi, 1999). Here, MERIS data are simulated by calculating the mean radiance of the AISA channels that are within a single MERIS channel. The algorithms were derived by choosing the MERIS channels that are the closest to the AISA channels used earlier (e.g., 521 nm becomes the channel centered at 510 nm, 700 nm becomes the channel centered at 705 nm, and so on). Because it is possible to find MERIS channels that are very close to the AISA channels, the resulting regression coefficients have about the same values as those derived with AISA data.

Discussion

The airborne water quality classification system was able to classify the target lakes with good accuracy despite different measurement configurations and lake types. This indicates that remote sensing is a useful tool for water quality classification. However, airborne remote sensing is quite expensive and its use will be limited to operational monitoring of large areas. Fortunately, the simulated Envisat MERIS data also gave good results.

For satellite instruments, atmospheric correction is more important than for airborne instruments because the radiance originating from below the water surface is very weak compared to the radiance from the atmosphere. This may reduce the estimation accuracy when satellite data is used instead of airborne data. On the other hand, the measurement conditions (e.g., solar angle, weather) will be more constant as the image is acquired in a single moment, which should improve the retrieval accuracy. Using retrieval algorithms based on channel ratio or difference indices reduces the effect caused by the atmosphere, but some kind of correction may still be necessary. For MERIS data, possible atmospheric correction methods are presented by Antoine and Morel (1999) and Moore et al. (1999).

The current operative lake classification system used in Finland is based on the measurements at fixed stations (laboratory analyses of water samples). These stations (or in some cases, just one station per lake) may not always present the actual condition of a lake in the best possible way. Perhaps the worst flaw of the current classification system is that the spatial resolution is limited. With remote sensing instruments, it is possible to see how the values of water quality variables are distributed spatially and thus get information on the complete status of the lake. Information on the relative spatial variations of water quality variables is also interesting, even though the absolute accuracy is not as good as with laboratory techniques.

The accuracy of a classification system also depends on the number of classes the system uses. In this analysis, the number of classes is only five or fewer, and part of the success may be attributed to that. However, in most cases, no information at all is available from smaller lakes, so even a coarse classification is useful. Furthermore, the experts who generated the operative classification systems discussed here have only used at most five classes.

One problem with low- and medium-resolution satellite data (e.g., MERIS) is that Finnish lakes are typically small and irregular in shape and may include small islands. The radiation reflected from the shore and the vegetation near the shore is usually stronger than the radiation from water. Therefore, if even a small portion of a pixel is covered by land, the retrieval of water quality variables may not be possible. However, the 300-m nadir resolution of MERIS should be good enough for large and medium size lakes if the rectification accuracy is good.

Case Study 2: Mapping Submerged Aquatic Vegetation

For many years, vegetation patterns have been studied and mapped by scientists and compared with previously mapped areas. As changes in the growth pattern become apparent, scientists can study the environmental influences that are causing these changes and determine if the changes are having a positive, negative, or neutral influence on neighboring ecosystems. Using hyperspectral remote sensing to study submerged aquatic vegetation gives researchers the chance to study and map larger areas at greater frequencies so that more complex and detailed conclusions about the state of the environment can be made.

Introduction

In 2003, Williams et al. investigated the effectiveness of using high spatial resolution hyperspectral remote sensing to map the distribution and abundance of submerged aquatic vegetation (SAV). In the same study, they investigated whether or not SAV could be mapped to the species level using this type of data.

SAV species differ in their tolerance to environmental factors. A shift in conditions of carbon availability, water clarity, or salinity, for example, could selectively affect the abundance of one species more than another. The presence, absence, and diversity of emergent and submerged aquatic species can be used to assess stream water quality and to rate stream degradation (Small et al., 1996).

Before changes in the distribution of species of SAV can be determined, however, an adequate species mapping technique must be developed. Once the species are mapped, causal relationships can be determined between environmental factors and changes in species coverage and distribution. Although aerial photographs are very useful for SAV abundance mapping, the lack of multispectral information in these photographs makes this data inadequate for species determination.

Williams et al. used hyperspectral imagery to identify and classify SAV beds in an aquatic environment that was characterized as optically complex given the significant concentrations of suspended solids and chlorophyll. The primary absorption band for photosynthesis (680 nm) was detectable in the submerged plant canopies. The differentiation of SAV species was done by exploiting the way light is scattered or absorbed by physically different plant canopies, rather than by some unique biochemical signature.

The data from Williams et al. suggest that the presence of epiphytes and sediment coating on the SAV obscure the biochemical reflectance signatures of the species. This biophysical methodology might be limited in beds of SAV plants with similar canopy profiles or during low tide when the leaves of meadow-forming species become horizontal at the surface.

Material and Methods

Study Site

The study area is part of the transition zone between the freshwater tidal river and the Chesapeake Bay estuary. The salinity is classified as oligo-haline (0.5 to 5.0 ppt). SAV species present at this site are primarily *Vallisneria americana* (wild celery) and *Myriophyllum spicatum* (Eurasian watermilfoil).

Field Spectroscopy

Ground-based in-situ spectra were obtained using an Applied Spectral Devices FR portable field spectrometer. Radiance and reflectance data for sample plots in Nanjemoy Creek and the Potomac River were obtained on 13 October 2000, by deploying a fiber-optic sensor head over beds of both milfoil and wild celery approximately 1 meter above the water surface. Spectra for material having a uniformly high reflecting spectral response such as beach sand were used to calibrate the airborne reflectance data for quality assurance. Laboratory spectra for field-collected milfoil and attached epiphyte colonies were obtained on 9 March 2001. Collected milfoil and wild celery, and calibration site spectra were entered into a spectral library data-base developed in MATLAB.

Hyperspectral Data and Image Processing

Airborne remotely sensed hyperspectral imagery for the site was acquired on 21 October 2000, using the HyMap system (Cocks et al., 1998). The flight-line dimensions were 2.3 × 20 km, and the ground sampling distance (pixel size) of the imagery was 4 m. Sensor radiance data were converted to appar-ent reflectance using ACORN, an atmospheric correction code based on the MODTRAN 4 radiative transfer model (ImSpec, LLC). Field sample plots were located in the HyMap imagery, and spectral signatures of SAV were extracted by averaging over a 50 pixel (200 m^2) area of interest for each plot.

A spectral transformation of the reflectance data was accomplished using continuum removal to plot the absorption bands at each wavelength. This procedure isolated the absorption band center and allowed these features to be easily compared with other reflectance spectra (Clark, 1999; Clark and Roush, 1984; Kruse et al., 1993). The depth of the absorption feature at a specific wavelength was used to identify the two species of SAV.

The first step in identifying the species of SAV in the imagery was to suppress the contributions of the optically active components in the ambient water, such as chlorophyll and free-floating algae. These components have spectral features that are similar to SAV in certain wavelengths.

Using the spectral signatures of the SAV species and the ambient water, a band math algorithm was developed to exploit the spectral differences of SAV versus ambient water at critical wavelengths. The algorithm first processed out the influence of the ambient water by using the continuum-removed spectral data. Band differencing was used to set any pixel that did not have absorption features associated with SAV to zero. A band ratio using two SAV absorption bands was then used to map the SAV beds. Because milfoil absorbs more strongly at the 681 nm band than wild celery, the band ratio was set up to take advantage of this difference along with another ratio for the 590 nm band as follows:

$$(\text{band } 1 - \text{band } 2) \times [(\text{band } 1/\text{band } 2) + (\text{band } 1/\text{band } 3)]$$

where band 1 = 604 nm, band 2 = 590 nm, band 3 = 681 nm.

This equation was used to segment the image and remove potential false positives. This preprocessing step also increased the speed of the next procedure by reducing the amount of data to be processed.

Pixels that scored in a set threshold were then passed to a SFF (spectral feature futting) procedure (ENVI, 1999) for SAV species identification. SFF is an algorithm that compares image spectral data to a set of reference spectra, in this case, the field-measured spectral library database, by a least-squares fit of the continuum-removed spectra (Clark et al., 1999). The spectral library database of field-collected spectra of milfoil and wild celery was compared to each pixel in the hyperspectral image by the SFF procedure.

The algorithm produced two images, a scale image measuring the depth of the absorption feature of interest, and a root-mean-square (rms) error image that indicates the degree of match between the reference spectra from the spectral database to the image spectra. Both images were then used to identify SAV by "best match" to the reference spectra, resulting in a determination of dominant SAV species in each target pixel.

Results and Discussion

Hyperspectral Imagery Interpretation

SAV beds were present and datable in the airborne hyperspectral imagery of Blossom Point. The two species of SAV and water were found to be spectrally

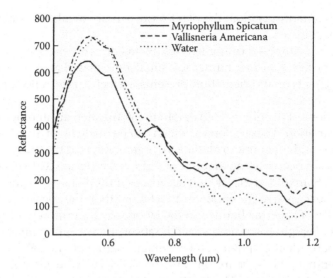

Figure 8.2 Relative reflectance spectra of two species of SAV and ambient water. (Williams D. J., N. B. Rybicki, A. V. Lombana, T. M. O'Brien, and R. B. Gomez. 2003. *Environmental Monitoring and Assessment*, 81(1–3), 383–392. With permission.)

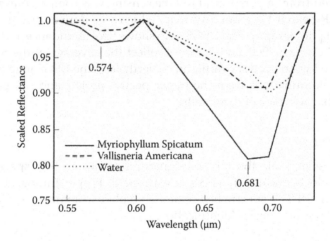

Figure 8.3 Continuum removed spectra of water and two species of SAV. (Williams D. J., N. B. Rybicki, A. V. Lombana, T. M. O'Brien, and R. B. Gomez. 2003. *Environmental Monitoring and Assessment*, 81(1–3), 383–392. With permission.)

separable (Figure 8.2). The absorption band depths at 681 nm and to a lesser extent 574 nm were more pronounced for milfoil than wild celery (Figure 8.3). This difference was likely due to the way the SAV plant canopy interacts with light (Asner, 1998).

The fully submerged profile of wild celery can be characterized as being meadow forming. Plant stems are mostly vertical. If the remote sensor is oriented at a nadir position, the plant tissue surface available as a reflecting target is negligible.

Milfoil is a canopy-forming species. Stems are vertical near the bottom, but this species also has numerous small branches that form horizontal surfaces. This type of orientation presents a much larger reflectance target relative to the sensor.

Interactions of the plants with epiphyte colonization are another source of spectral variation. These colonies, which can be made up of algae, sediment, bryophytes, and other micro and macro organisms, coat the SAV leaf surface and therefore decrease the amount of light reaching the leaf surface (Orth et al., 1982; Stankelis et al., 2000). The species milfoil, due to its morphology, has more surface area for epiphyte attachment than the species wild celery. Comparison of spectra obtained in the laboratory for milfoil with and without attached epiphytes indicates that the absorption at approximately 574 nm is mostly a result of epiphytes and sediment coating on the SAV.

Two remnant beds of milfoil and wild celery were identified by the hyperspectral technique. The locations of the beds correspond well to the locations of the bed during the peak growing season determined by the U.S. Geological Survey (USGS). The accuracy of the SAV species map that was derived from hyperspectral remote sensing was compared to the USGS National Research Program SAV map, which was derived using field data and aerial photography. The USGS map shows SAV abundance and distribution on 31 August 2000, and does not reflect the coverage of the SAV at the time of the hyperspectral overflight. Nevertheless, the USGS map reinforces species determinations from the hyperspectral project and can be used to estimate the accuracy of the results.

Conclusion

Further research into the spectral signatures of various SAV species with a focus on biochemical differences is warranted. Proper timing of the overflight to collect SAV spectra before dense epiphyte colonization might allow for more accurate species identification.

References

Antoine, D., and A. Morel. 1999. A multiple scattering algorithm for atmospheric correction of remotely sensed ocean colour (MERIS instrument): Principle and implementation for atmospheres carrying various aerosols including absorbing ones. *International Journal of Remote Sensing*, 20 (9), 1875–1916.

Asner, G. P. 1998. Biophysical and biochemical sources of variability in canopy reflectance, *Remote Sens. Environ.* 64, 234–253.

Clark, R. N. 1999. Spectroscopy of rocks and minerals, and principles of spectroscopy, In: *The Manual of Remote Sensing*, 3rd Edition (A Series): Volume 3: *Remote Sensing for the Earth Sciences*,. A. Rencz, (ed.), John Wiley and Sons, New York.

Clark, R. N., and T. L. Roush. 1984. Reflectance spectroscopy: Quantitative analysis techniques for remote sensing applications, *Journal of Geophysical Research*, 89(B7), 6329–6340.

Cocks, T., R. Jenssen, A. Stewart, I. Wilson, and T. Shields. 1998. The HyMap airborne hyperspectral sensor: The system, calibration and performance, Proceedings of the 1st EARSEL Workshop on Imaging Spectroscopy, Zurich, October.

Dekker, A. G. 1993. Detection of Optical Water Parameters for Eutrophic Lakes by High Resolution Remote Sensing. PhD Dissertation, Free University, Amsterdam.

Dekker, A. G., T. J. Malthus, and E. Seyhan. 1991. Quantitative modeling of inland water quality for high-resolution MSS systems, *IEEE Transactions on Geoscience and Remote Sensing*, 29 (1), 89–95.

ENVI. 1999. *ENVI Users Guide*, Research Systems, Boulder, Colorado.

Gitelson, A. 1992. The peak near 700 nm on radiance spectra of algae and water: Relationships of its magnitude and position with chlorophyll concentration. *International Journal of Remote Sensing*, 13, 3367–3373.

Gitelson, A., G. Garbuzov, F. Szilagyi, K. H. Mittenwey, K. Karnieli, and A. Kaiser. 1993. Quantitative remote sensing methods for real-time monitoring of inland water quality. *International Journal of Remote Sensing*, 14, 1269–1295.

Kallio, K., T. Kutser, T. Hannonen, S. Koponen, J. Pulliainen, J. Vepsalainen, and T. Pyhalahti. 2001. Retrieval of water quality variables from airborne spectrometer in various lake types at different seasons. *The Science of the Total Environment*, 268(1–3), 59–78.

Koponen S., J. Pulliainen, K. Kallio, M. Hallikainen. 2002. Lake water quality classification with airborne hyperspectral spectrometer and simulated MERIS data, *Remote Sensing of Environment*, 79(1), 51–59.

Koponen, S., J. Pulliainen, H. Servomaa, Y. Zhang, M. Hallikainen, K. Kallio, K. Eloheimo, and T. Hannonen. 2001. Analysis on the feasibility of multi-source remote sensing observations for chl-*a* monitoring. *The Science of the Total Environment*, 268(1–3), 95–106.

Kruse, F. A., A. B. Lefkoff, and J. B. Dietze. 1993. Expert system-based mineral mapping in northern Death Valley, California/Nevada using the Airborne Visible/Infrared Imaging Spectrometer (AVIRIS), *Remote Sens. Environ.* 44, 309–336.

Makisaru, K., M Meinander, M. Rantasuo, J. Okkonen,M. Aikio, K. Sipola, P. Pylkko, and B. Braam. 1993. Airborne Imaging Spectrometer for Applications (AISA). *Digest of IGARSS '93*, Tokyo, Japan, August 18–21, 2, 479–481.

Moore, G., J. Aiken, and S. Lavender. 1999. The atmospheric correction of water colour and the quantitative retrieval of suspended particulate matter in Case II waters: Application to MERIS, *International Journal of Remote Sensing*, 20 (9), 1713–1733.

Morel, A., and L. Prieur. 1977. Analysis of variations in ocean color, *Limnology and Oceanography*, 22, 709–722.

Murphy, J., and J. P. Riley. 1962. A modified single solution method for the determination of phosphate in natural waters. *Analytica Chimica Acta*, 27, 31–36.

Orth, R. J., K. A. Moore, and J. van Montfrans. 1982. Submerged aquatic vegetation: Distribution and abundance in the lower Chesapeake Bay and interactive effects of light, epiphytes and grazers, Final Report, USEPA Chesapeake Program Grant No. X003246, 141 pp.

Preisendorfer, R. W. 1976. *Hydrologic optics: Vol. 1. Introduction*. Honolulu: U.S. Department of Commerce National Oceanic and Atmospheric Administration, Environment Research Laboratory, 218 pp.

Pulliainen, J., K. Kallio, K. Eloheimo, S. Koponen, H. Servomaa, T. Hannonen, S. Tauriainen, and M. Hallikainen. 2001. A semi-operative approach to water quality retrieval from remote sensing data, *The Science of the Total Environment*, 268(1–3), 79–94.

Rast, M., J. Bezy, and S. Bruzzi. 1999. The ESA Medium Resolution Imaging Spectrometer MERIS— A review of the instrument and its mission, *International Journal of Remote Sensing*, 20(9), 1681–1702.

Small, A. M., D. L. Roberts, W. H. Adey, S. M. Lutz, and E. G. Reese. 1996. A macrophyte-based rapid biosurvey of stream water quality: Restoration at the watershed scale, *Restoration Eco.* 4(2), 124–145.

Smith, R. C. and K. S. Baker. 1978. Optical classification of natural waters, *Limnology and Oceanography*, 23, 260–267.

Stankelis, R. M., W. R. Boynton, and J. M. Frank. 2000. Submerged aquatic vegetation (SAV) habitat evaluation, In *Chesapeake Bay Water Quality Monitoring Program Ecosystem Processes Component (EPC) Level One Report #17* (interpretive). University of Maryland Center for Environmental Science, Technical Report Series No. TS-252-00-CBL, Ref. No. [UMCES] CBL 00-0174.

Williams D. J., N. B. Rybicki, A. V. Lombana, T. M. O'Brien, and R. B. Gomez. 2003. Preliminary investigation of submerged aquatic vegetation mapping using hyperspectral remote sensing, *Environmental Monitoring and Assessment*, 81(1–3), 383–392.

9

Forestry Applications

Foresters who are responsible for maintaining the health and viability of our forests rely on early detection schemes to let them know when a problem may be arising. To determine changes in the condition and amount of the vegetation, remote sensing imagery from airborne and satellite-based sensors has been used to map forested areas. Forest stands experiencing higher stress can then be examined on the ground to identify the causes of the stress (e.g., beetle attack, root rot, poor site conditions).

The two studies below discuss the use of hyperspectral imagery to study and map two indicators of forest health: insect infestation and chlorophyll content. In the first study, Lawrence and Labus (2003) used hyperspectral imagery to detect a Douglas-fir beetle infestation in its early stages, and in the second study, Sampson et al. (2003) used hyperspectral imagery to estimate the chlorophyll content in tolerant hardwoods.

Using the information from these types of studies aids researchers in tracking and treating potential problems before significant areas of the forest are destroyed. Ameliorative actions, if available, could be taken at earlier stages, which would reduce the adverse economic and forest health effects.

Case Study 1: Detecting Douglas-Fir Beetle Infestation

Introduction

Early detection of insect infestations and forest diseases, such as beetle or root rot, is important to foresters who want to minimize economic loss due to these threats (Schmitz and Gibson, 1996). In rugged terrain where Douglas fir (*Pseudotsuga menziesii*) often grows, this monitoring typically requires extensive work in the field that is both time consuming and expensive.

In the late 1990s, remote sensing imagery from airborne and satellite-based sensors was used to map infestations. Unfortunately, this imagery lacked the spectral sensitivity to detect the problem before visual signs of the infestation become evident. Although newer hyperspectral instruments have this required sensitivity and can provide information comparable to spectra obtained in the laboratory, these instruments have lacked the spatial resolution to map individual tree canopies.

When commercial high-resolution, hyperspectral imagery became available, Lawrence and Labus (2003) conducted an assessment of tree stress to determine if they could detect the early stages of infestation or disease over large areas more quickly and efficiently than by ground observations.

Using the newest commercially available hyperspectral imagery, Lawrence and Labus extended the studies done by previous researchers to detect early stress caused by Douglas-fir beetle (*Dendroctonus pseudotsugae*) infestation at the individual tree or subcanopy level.

Insects, such as the Douglas-fir beetle, are considered agents of stress in forests because they adversely affect the physiology and growth of trees, often killing them. The Douglas-fir beetle occurs throughout much of the western United States, British Columbia, and Mexico (Schmitz and Gibson, 1996; Thomson et al., 1996). These beetles normally attack and kill small groups of trees, but during outbreaks, attacks on tree groups as large as 100 are not uncommon, especially in dense stands.

Early evidence of infestation consists of entry holes in the tree bark and frass expelled from bark crevices by invading beetles. Several months after a successful infestation, foliage exhibits chlorosis by turning yellow, then sorrel, and then reddish brown, with needles beginning to fall from infested trees the year following the attack. These changes in leaf physiology, chemistry, and photosynthetic efficiency affect the reflectance response of vegetation (Sampson et al., 1998). The detailed shape of the reflectance spectra and variables such as width, depth, skewness, and symmetry of absorption features can be measured and used to detect canopy stresses.

For example, leaf pigments, chlorophyll a and b, and chlorophyll fluorescence levels in leaves and needles of trees are highly related to visible and near-infrared ratios and indices, particularly red-edge indices, at the leaf and simulated canopy level (Sampson et al., 1998; Zarco-Tejada et al., 1999). Red-edge indices are calculated along the red/infrared boundary, where chlorophyll absorption in vegetation forms one of the most extreme slopes found in spectra of naturally occurring materials. In healthy, green vegetation, the edge is sharp and steep, but as vegetation becomes stressed or senescence starts, the width of the absorption band decreases, and the red edge shifts toward shorter wavelengths (Clark et al., 1995).

Methods

A Douglas-fir stand that was infested with the Douglas-fir beetle was located in the Lamar Valley of Yellowstone National Park, Wyoming. Sampled trees were selected randomly from 1:5,000 scale color infrared aerial photos. These trees were then clustered into tree health classes based on field observations and were grouped as:

1. Healthy (H)—no sign of beetle infestation or other damage.
2. Attacked (A)—beetle infestation present as evidenced in bark, but the tree crown remained green with no visual signs of decline.
3. Dead (D)—successful beetle infestation that has killed the tree within the past year, evidenced by red or yellow foliage.

All sampled trees were dominant or subdominant with a diameter ranging from 0.5 to 2 m at breast height.

A hyperspectral image swath of the sampled stand was collected at 10 a.m. on 4 August 1999, from the Probe-1 sensor. The sensor was flown aboard a helicopter at 500 m, producing 1-m² pixel size with an approximately 0.5-km swath width. The sensor collected 128 continuous spectral bands in the visible through SWIR spectral regions (0.4 to 2.5 μm). Spectral responses were also extracted for other cover types within the study areas, including light yellow (senescent) grass (LG) (8 pixels), heavy green grass (HG) (18 pixels), and shadow (SH) (18 pixels), to differentiate these spectral responses from the trees.

Individual and class-average spectra were plotted for visual examination of among-class separability. Of several analysis methods examined, the two best performing methods were stepwise discriminant analysis (DISCRIM) and classification and regression tree analysis (CART).

Results

Examination of spectral responses from individual trees showed classes grouping at different reflectance values in specific wavelengths, such as in the two sharp peaks at 1,000 and 1,100 nm, the wider peak around 1,250 nm, and the two large peaks at <1,500 nm (Figure 9.1). There were also considerable

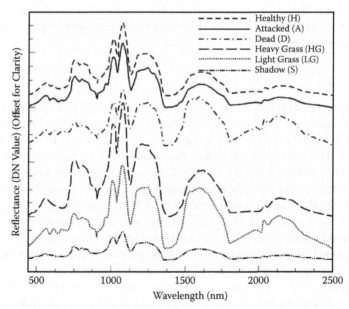

Figure 9.1 Spectral responses at different reflectance values in specific wavelengths. (Lawrence, R., and M. Labus. 2003. *Western Journal of Applied Forestry*, 18(3), 202–206. With permission.)

overlaps among different classes in certain portions of the spectrum, even within those regions where some class separability existed. For example, only class SH separated well from the other classes in all portions of the spectrum due to the very low reflectance of shadow.

In the visible wavelengths, no vegetation class showed substantial separability. At the NIR (near infrared) boundary (700–750 nm) and in the short-wave IR region (especially 1,100–1,200 nm) where vegetation characteristics typically stand out, the D class was differentiated from the spectra of other green vegetation, but the H and A classes failed to differentiate. HG was separated from other classes with very high reflectance values within the 750–800-nm and 1,000–1,100-nm range, was strongly mixed with other spectra in the 1,500–1,750-nm range, and then again separated into its own group in the 2,000–2,500-nm range, although with intermediate reflectance values at this wavelength. The H and D classes separated well in the 1,500–1,750-nm and 2,000–2,500-nm ranges, while the A class spectra overlapped the H class in these wavelengths.

An examination of average class spectra gave a clearer overview of class separability (Figure 9.1). Green vegetation separated from the D class in the peaks of the 400–800-nm and the 1,000–1,375-nm ranges. In the visible range, all classes except shadow were confounded. Two sharp reflectance peaks at 1,007 and 1,069 nm showed promise for good class separability in all classes. At longer wavelengths (1,500–2,500 nm), grasses (heavy and light) were confounded with the D class. In addition, the averaged A spectrum was very similar to the H class, indicating poor separability for this important class at these longer wavelengths.

Discussion

The analysis demonstrated that subcanopy resolution hyperspectral imagery could successfully distinguish among tree stress classes resulting from Douglas-fir beetle attack. Examination of the classification methods showed that the CART approach provided the best ability to separate tree health classes. The ability of CART to use different band combinations for each class in a rule-based classification allowed for maximum spectral separability of tree health classes compared to DISCRIM, which required the same spectral bands for all classes. This ability was advantageous because tree health classes and other background classes were different in their physical and chemical characteristics, and thus spectral regions in which classes could be distinguished varied.

The main spectral profiles for each class showed these class spectra and the spectral regions where class distinctions could be made. Although the spectra were similar, slight shifts in the spectral regions of maximum separation could be seen in all classes. CART used these slight differences in the spectra to build the classification tree, thus taking full advantage of the spectral resolution afforded by hyperspectral imagery and reducing the spectral data to those bands that provided the best class distinctions.

This study showed that CART analysis of remote sensing data is a robust and easily implemented statistical method of classification without the need of extensive expert knowledge. Because this study was conducted in a single stand having one species and a single known source of tree stress, further studies will be necessary to determine how broadly the results are applicable. At least in this case, however, CART effectively created classification rules that distinguished the early stages of tree stress by using the full spectral capabilities of hyperspectral imagery. Thus, hyperspectral imagery of large forest stands might be useful in identifying areas of relative tree stress that would not otherwise be detected without prohibitive field reconnaissance.

Case Study 2: Estimating Chlorophyll Content in Tolerant Hardwoods

Introduction

The second case study for this chapter was conducted by Sampson et al. (2003), who used remote hyperspectral data to estimate the amount and type of chlorophyll content in hardwood trees. Developing spectral features related to chlorophyll or other pigments is useful in identifying whether forests are healthy or are stressed to the point where productivity of the resource may decrease.

The traditional method of assessing the health of a tree, a visual inspection, is subjective and does not directly measure tree vigor (Ferretti, 1997). In contrast, a nonvisual method that allows tracking of pigment concentrations (e.g., chlorophyll) could provide an objective, early warning indicator of stand condition. Early detection could help to identify stands requiring remedial or salvage action before damage is visible and potentially before biomass loss occurs.

Optical indices derived from the red edge (the region of rapid transition between red and near infrared reflectance) are especially useful because they are sensitive to both chlorophyll content (chl_{a+b}) and canopy structure. Several investigators have related changes in chl_{a+b} to a shift in position of the spectral red edge (e.g., Horler et al., 1983; Vogelmann et al., 1993; and Gitelson et al., 1996). This shift has been associated with plant stress, forest decline, and leaf development (e.g., Rock et al., 1988; Boochs et al., 1990; Miller et al., 1991; Hoque and Hutzler, 1992).

Stress can affect other physiological features such as leaf water content. However, changes in leaf water content are less sensitive than those of chl_{a+b} because they are measurable only under severe dehydration events (Carter, 1993).

Nitrogen deficiency is the second greatest factor limiting tree growth, next to water stress (Kramer and Kozlowski, 1979). Foliar chl$_{a+b}$ has been positively correlated with foliar nitrogen (N) in western red cedar (*Thuja plicata* D.) (Radwan and Harrington, 1986), big leaf maple (*Acer macrophyllum* P.) (Yoder and Pettigrew-Crosby, 1995), sugar maple (*Acer saccharum* M.) (Ellsworth, 1999), and balsam fir (*Abies balsamea* L.) (Luther and Carrol, 1999).

Foliar chl$_{a+b}$ has also been positively correlated with shoot growth rate in western red cedar (Radwan and Harrington, 1986) and balsam fir (Luther and Carrol, 1999) and with photosynthetic rate in sugar maple (Ellsworth, 1999). Therefore, foliar chl$_{a+b}$ is likely a sensitive indicator of tree physiological condition.

Advances in technology, data processing, and scientific application of findings made remote sensing approaches more practical. Airborne hyperspectral technologies themselves progressed markedly, offering improved data capture and processing capabilities along with fine spectral and spatial detail. For example, the Compact Airborne Spectrographic Imager (CASI) was used in various forestry applications, which include conducting land classification (e.g., Zarco-Tejada and Miller, 1999), inventorying forests (e.g., Davison et al., 1999a, 1999b), assessing forest management practices (Sampson et al., 2001), detecting root rot (e.g., Reich and Price, 1999), and identifying insect damage (e.g., Leckie et al., 1989).

For the case study discussed below, Sampson et al. (2003) examined the use of CASI technology to estimate chl$_{a+b}$ in a managed tolerant hardwood forest in the Algoma Region of Ontario, Canada. One objective was to determine if chlorophyll content could be predicted following different harvesting practices. The second objective was to estimate chl$_{a+b}$ across seasons (i.e., conduct change detection analysis) on a range of maple sites. The overall aim was to develop a prototype system for monitoring forest physiological condition.

Discussion

Turkey Lakes Harvesting Impacts Project

Interpreting the findings of the Sampson et al. (2003) Turkey Lakes Harvesting Impacts Project (TLHIP) study requires some understanding of the likely factors affecting chl$_{a+b}$ estimation. For instance, the wide range of LAI (leaf area index) values (from 1.1 to 4.2) and the fairly narrow range of chl$_{a+b}$ (from 26 to 56 μg/cm^2) were likely reasons for the Root Mean Square Error (RMSE) being higher and the correlations lower than those found in a previous study using the same approach for sugar maple (Zarco-Tejada et al., 2001). When tree crowns are not dense (e.g., LAI values are less than 2), the spectral effect of the underlying soil and vegetation can mask the condition of foliage (Guyot et al., 1989).

Canopy geometry also influences the ratio of shadowed surfaces to illuminated surfaces (i.e., bidirectional aspects). Zarco-Tejada et al. (2001) showed that chlorophyll estimates in closed maple stands (i.e., LAI greater than two) are not significantly affected by shadowed components when appropriate inversion methods are used. However, other species with open canopies and/or different canopy architectures (e.g., conical shape) could influence these predictions. To address these potential challenges, researchers are studying optimum view angles in different species and collecting airborne and field data on well-characterized sites using optically based methods (e.g., Chen, 1996).

Spectral Indices

Another challenge in estimating chl_{a+b} is choosing a spectral index. Evidence from both leaf- and canopy-scale experiments demonstrate that relationships exist between pigment concentrations and narrow band reflectance, but many indices developed at the leaf level do not work at the canopy level because the leaf and canopy media have different optical properties. The large number of optical indices developed at the leaf level also makes it difficult to decide which index to use (Blackburn, 1999).

Most proposed spectral indices have been developed empirically, typically by combining bands that are sensitive and insensitive to stress. The advantage of this approach is that insensitive bands function as baselines that factor out variability due to causes other than variation in leaf chl_{a+b}. Similarly, spectral derivatives have important advantages over spectral reflectance; for example, they can reduce variability due to changes in illumination or background reflectance (Elvidge and Chen, 1995). Myneni et al. (1995) quantified possible physical reasons for these observed correlations.

For this Sampson et al. (2003) case study, the researchers examined several indices for estimating chlorophyll in sugar maple at the leaf and canopy levels (Zarco-Tejada et al., 2001) and selected the spectral index (R_{750}/R_{705}). Widely used indices for pigment estimation, such as the Normalized Difference Vegetation Index (NDVI), primarily track canopy structural changes and are therefore considered insensitive to subtle changes in pigment content. The spectral index (R_{750}/R_{705}), on the other hand, offers a means to track chl_{a+b} semi-empirically (i.e., without detailed modeling) or predictively using model inversion.

Spatial Resolution

A final consideration in estimating chl_{a+b} is defining the optimal, spectral, and temporal resolutions because these resolutions are generally not known (Treitz and Howarth, 2000). This issue can be approached from two directions. The more common approach is to use the spatial resolution that will capture the spatial variability of the scene to be used. The spatial variability is a function of the type of environment being studied and the type of information required (Woodcock and Strahler, 1987). But selecting an appropriate scale for vast forested areas is a challenge; Ontario alone has 39 million ha of productive forest. This challenge was illustrated by Marceau et al. (1994) who examined natural forest environments at various scales and concluded that there is no unique spatial resolution at which all geographic entities could be discriminated.

A second, equally important approach is to choose a spatial resolution that ensures that the assumptions made in the remote sensing interpretation methodologies are valid. For example, at very high spatial resolution, crown reflectance and canopy reflectance must be distinguished in parameter retrieval results.

In an earlier study, Sampson et al. (2001) showed that an appropriate resolution (as derived by semivariogram analysis) is either approximately 5 m or approximately 9 m, depending on whether physiological or structural parameters are being examined. Similarly, optimal spatial resolutions for several boreal forest species were shown to be in the order of 3 to 7 m (Treitz and Howarth, 2000). Extending these suggested resolutions to other environments may not be possible given the potential influences of factors such as terrain and sensor characteristics. However, the results do provide some guidance in considering the suitability of existing data sets and in developing sampling strategies in these forest types.

Chlorophyll Content (chl$_{a+b}$)

Chlorophyll content, including the ratio of chlorophyll *a/b*, could be affected by a range of intrinsic and extrinsic factors and is subject to considerable natural and stress-induced variation (Kozlowski and Pallardy, 1997a). For instance, chlorophyll content is higher in sun-adapted foliage than in shade leaves. Chlorophyll content is also influenced by genetic factors, which produce plants ranging from albinos, devoid of chlorophyll, to those showing various degrees of striping or mottling. Flooding or drought may induce chlorosis, and chlorophyll synthesis may be inhibited by leaf diseases. Chlorophyll content is also affected by seasonal cycles (Vogg et al., 1998; Rosenthal and Camm, 1997) and by extreme air and soil temperatures (Kozlowski and Pallardy, 1997b). In addition, both chronic and acute injury from atmospheric pollutants may be manifested as chlorosis and then leaf senescence with symptoms more conspicuous in angiosperms than gymnosperms, in seedlings rather than older trees, and in immature rather than mature foliage (Treshow and Pack, 1970).

Because of its nonspecific nature, reflectance should be used with caution when attempting to diagnose a stress. Reflectance could, however, be combined with other analytical methods to pinpoint causal factors. For instance, the decline phenomenon in sugar maple found in Canada and the United States has been often been correlated with nutrient stress, especially calcium (Ca) and magnesium (Mg) (Bernier and Brazeau, 1988; Horsley et al., 2000).

The results from this case study (Sampson et al., 2003) show positive correlation (significant in all but one case) between foliar chl$_{a+b}$ and Ca and

Mg content of leaves in both the change analysis sites and the TLHIP plots. Thus, the differences in chlorophyll content appear to relate somewhat to the nutrient stress status of the sugar maple stands. Incorporating spatial data such as soil type, terrain, and insect or disease surveys could offer greater insight into why chlorophyll changes are evident.

Monitoring Approach

The wide variety of factors producing chlorosis suggests that chlorosis could be caused by general disturbances of metabolism, as well as by deficiency (or toxicity) of a specific mineral element (Kozlowski and Pallardy, 1997a). To account for these factors, a diagnostic approach could be used to integrate various types of information (e.g., climatic, terrain, and vegetation) with data from sampling chlorophyll over several seasons to assess trends. A similar strategy would be to target a specific pest problem (such as root rot disease, e.g., *Armillaria ostoyae*) that causes chlorosis, and then to allow its spread and extent to be monitored. Alternatively, with a bioindicator approach (e.g., Sampson et al., 2000), the causal agents influencing chlorosis are not directly of interest, but rather attention is placed on identifying deviations from the "normal" chlorophyll range. All these strategies have merit, and the choice depends on management objectives and overall feasibility.

In this case study (Sampson et al., 2003), samples were collected under full leaf expansion prior to senescence on sites that included many scientific interests. The rationale was to minimize early- and late-season phenological changes that influence spectral response while targeting sites of long-term research importance. In addition, stand level assessments were emphasized because, as described by others (e.g., Mohammed et al., 1997; Colombo and Parker, 1999):

- Stands are the basic operational management units in a forest ecosystem.
- Stands have some uniformity of tree species and age, making it easier to compare physiological attributes among similar ecosystem units.
- Stands provide a means to examine spatial variability of larger land units such as the landscape.

As satellite sensors that could estimate chl_{a+b} become available [e.g., MERIS, (Moderate-resolution Imaging Spectroradiometer) MODIS], target stands may still be used, but it may also be possible to explore regional patterns of variation. The red-edge index (R_{750}/R_{705}) will be useful because it provides a link between satellite and airborne platforms. This robust measure of chl_{a+b} minimizes the confounding influences of structure in tolerant hardwoods yet is sensitive to subtle changes in optical properties at both the leaf and canopy levels.

Conclusion

Predicting chl_{a+b} or any other canopy biophysical parameter from airborne or satellite canopy reflectance is essential to many aspects of forest management. Although statistical relationships have been widely used in estimating biophysical features, no predictive estimates can be inferred from these studies because these locally derived relationships are affected by species, canopy structure, and solar-viewing geometry. To advance the predictive capability for chl_{a+b}, radiative transfer (RT) simulation for leaf and canopy levels with red-edge indices such as R_{750}/R_{710} and R_{750}/R_{705} can be used in models that are both predictive and simplistic.

The practical implications of these findings are in providing an early warning measure of condition without needing ground data collection. This approach has several advantages over traditional ground surveys, which primarily rely on structural measures and subjective vigor estimates.

Although the research findings of the Sampson et al. (2003) and Lawrence and Labus (2003) case studies are encouraging, the robustness of the algorithms and approaches needs to be addressed. In particular, applying this method in open canopies and for different species needs to be investigated. Moreover, if chlorophyll estimates and the CART classification method are to be used as monitoring tools, the factors affecting response need to be more fully understood.

The question of operational feasibility remains. The cost and limitations of airborne hyperspectral imagery must be considered; however, its utility increases when considering its broad applicability to many aspects of forest management, such as forest health monitoring and forest mapping and inventory. It would be helpful if hyperspectral data now obtained via aircraft became available at a reasonable cost via satellites. Ready access to satellite-based pigment estimates of a well-known scene would also help us answer questions about its potential, as well as factors limiting its retrieval.

References

Bernier, B., and M. Brazeau. 1988. Magnesium deficiency symptoms associated with sugar maple dieback in a Lower Laurentians site in southeastern Quebec, *Canadian Journal of Forest Research*, 18, 126–129.

Blackburn, G. A. 1999. Relationships between spectral reflectance and pigment concentrations in stacks of deciduous broadleaves, *Remote Sensing of Environment*, 70, 224–237.

Boochs, F., G. Kupfer, K. Dockter, and W. Kuhbauch. 1990. Shape of the red edge as vitality indicator for plants, *International Journal of Remote Sensing*, 11(10), 1741–1753.

Carter, G. A. 1993. Responses of leaf spectral reflectance to plant stress, *American Journal of Botany*, 80, 239–243.

Chen, J. M. 1996. Optically-based methods for measuring seasonal variation of leaf area index in boreal conifer stands, *Agricultural and Forest Meteorology*, 80, 135–163.

Clark, R. N., T. V. V. King, C. Ager, and G. A. Swatze. 1995. Initial vegetation species and senescence/stress indicator mapping in the San Luis Valley, Colorado using imaging spectrometer data, In: Proceedings of Summitville Forum '95, Posey, H. H., J. A. Pendelton, and D. Van Zyl (eds.). *Colorado Geological Survey Special Publication*, 38, 64–69.

Colombo, S. J., and W. C. Parker. 1999. Does Canadian forestry need physiology research? *Forestry Chronicle*, 75, 667–673.

Davison, D., S. Achal, S. Mah, R. Gauvin, M. Kerr, A. Tam, and S. Preiss. 1999a. Determination of tree species and tree stem densities in northern Ontario forests using airborne CASI data, In: Proceedings of the Fourth International Airborne Conference and Exhibition, Ottawa, ON, 187–196.

Davison, D., R. Price, S. Mah, R Gauvin, and S. Achal. 1999b. Forest analysis under partly cloudy conditions for large mosaics of CASI data, In: Proceedings of the Fourth International Airborne Conference and Exhibition, Ottawa, ON, 255–264.

Ellsworth, D. S. 1999. Nitrogen addition affects leaf nutrition and photosynthesis in sugar maple in a nutrient-poor northern Vermont forest, In: Sugar Maple Ecology and Health: Proceedings of an International Symposium General Technical Report NE-261. Radnor, PA: U.S. Department of Agriculture, Forest Service, Northeastern Research Station, 98–105.

Elvidge, C. D., and Z. Chen. 1995. Comparison of broad-band and narrow-band red and near-infrared vegetation indices, *Remote Sensing Environment*, 54, 38–48.

Ferritti, M. 1997. Forest health assessment and monitoring-issues for consideration, *Environmental Monitoring and Assessment*, 48, 45–72.

Gitelson, A. A., M. N. Merzlyak, and H. K. Lichtenthaler. 1996. Detection of red-edge position and chlorophyll content by reflectance measurements near 700 nm, *Journal of Plant Physiology*, 148, 501–508.

Guyot, G., D. Guyon, and J. Riom. 1989. Factors affecting the spectral response of forest canopies: A review, *Geocarto International*, 4(3), 3–18.

Holmgren, P., and T. Thuresson. 1998. Satellite remote sensing for forestry planning: A review, *Scandinavian Journal of Forest Research*, 13, 90–110.

Hoque, E., and P. J. S. Hutzler. 1992. Spectral blue-shift of red edge monitors damage class of beech trees, *Remote Sensing Environment*, 39, 81–84.

Horler, D. N., M. Dockray, and J. Barber. 1983. The red edge of plant leaf reflectance, *International Journal of Remote Sensing*, 4, 273–288.

Horsley, S. B., R. P. Long, S. W. Bailey, R. A. Hallet, and T. J. Hall. 2000. Factors associated with the decline disease of sugar maple on the Allegheny Plateau, *Canadian Journal of Forest Research*, 30, 1365–1378.

Kozlowski T. T., and S. G. Pallardy. 1997a. *Growth Control in Woody Plants*. Academic Press, Toronto, Ontario, Canada, 641 pp.

Kozlowski, T. T., and S. G. Pallardy. 1997b. *Physiology of Woody Plants, Second Edition*, Academic Press, Toronto, Ontario, Canada, 411 pp.

Kramer, P. J., and T. T. Kozlowski. 1979. *Physiology of Woody Plants*. Academic Press, New York, 811 pp.

Lawrence, R., and M. Labus. 2003. Early detection of Douglas-fir beetle infestation with subcanopy resolution hyperspectral imagery, *Western Journal of Applied Forestry*, 18(3), 202–206.

Leckie, D. G., D. P. Ostaff, P. M. Teillet, and G. Fedosejevs. 1989. Spectral characteristics of tree components of balsam fir and spruce damaged by spruce budworm, *Forest Science*, 35, 582–600.

Luther, J. E., and A. L. Carroll. 1999. Development of an index of balsam fir vigor by foliar spectral reflectance, *Remote Sensing Environment*, 69, 241–252.

Marceau, D. J., D. J. Gratton, R. A. Fournier, and I. P. Fortin. 1994. Remote sensing and the measurement of geographical entities in a forested environment, 2: The optimal spatial resolution, *Remote Sensing Environment*, 49, 105–117.

Miller, J. R, W. Jiyou, M. G. Boyer, M. Belanger, and E. W. Hare. 1991. Seasonal patterns in leaf reflectance red-edge characteristics, *International Journal of Remote Sensing*, 12(7), 1509–1523.

Mohammed, G. H, P. H. Sampson, S. J. Colombo, T. L. Noland, and J. R Miller. 1997. Bioindicators of forest sustainability: Development of a forest condition rating system for Ontario. Ontario Ministry of Natural Resources, Ontario Forest Research Institute, Sault Ste. Marie, ON. Forest Research Information Paper No.137. 22 pp.

Myneni, R. B., F. G. Hall, P. J. Sellers, and A. L. Marshak. 1995. The interpretation of spectral vegetation indexes, *IEEE Transactions on Geoscience & Remote Sensing*, 33, 481–486.

Radwan, M. A., and C. A. Harrington. 1986. Foliar chemical concentrations, growth, and site productivity relations in western red cedar, *Canadian Journal of Forest Research*, 16, 1069–1075.

Reich, R. W., and R. Price. 1999. Detection and classification of forest damage caused by tomentosus root rot using an airborne multispectral imager (CASI), In: Proceedings of International Forum: Automated interpretation of high spatial resolution digital imagery for forestry, Hill, D. A., and D. G. Leckie (eds.). Canadian Forest Service, Pacific Forestry Centre, Victoria, B.C., 179–185.

Rock, B. N., T. Hoshizaki, and J. R. Miller. 1988. Comparison of in situ and airborne spectral measurements of the blue shift associated with forest decline, *Remote Sensing Environment*, 24, 109–127.

Rosenthal S. I., and E. L. Camm. 1997. Photosynthetic decline and pigment loss during autumn foliar senescence in western larch (*Larix occidentalis*), *Tree Physiology*, 17, 767–775.

Sampson, P. H., G. H. Mohammed, P. J. Zarco-Tejada, J. R. Miller, T. L. Noland, D. Irving, P. M. Treitz, S. J. Colombo, and J. Freemantle. 2000. The bioindicators of forest condition project: A physiological, remote sensing approach, *The Forestry Chronicle*, 76(6), 941–952.

Sampson, P. H, P. J. Zarco-Tejada, G. H. Mohammed, J. R. Miller, and T. L. Noland. 2003. Hyperspectral remote sensing of forest condition: Estimating chlorophyll content in tolerant hardwoods, *Forest Science*, 49(3), 381–391.

Sampson, P. H., P. M. Treitz, and G. H. Mohammed. 2001. Forest condition assessment in tolerant hardwoods: An examination of spatial scale, structure and function, *Canadian Journal of Remote Sensing*, 27(3), 232–246.

Schmitz, R. F., and K. E. Gibson. 1996. Douglas-fir Beetle. USDA Forest Service Forest Insect & Disease Leaflet 5, R1-96-87, Washington, D.C.

Thomson, A. J., D. G. Goodenough, H. J. Barclay, Y. J. Lee, and R. N. Sturrock. 1996. Effects of laminated root rot (*Phellinus weirii*) on Douglas-fir foliar chemistry, *Canadian Journal of Forest Research*, 26, 1440–1445.

Treitz, P. M., and P. J. Howarth. 1999. Hyperspectral remote sensing for estimating biophysical parameters of forest ecosystems, *Progress in Physical Geography*, 23(3), 359–390.

Treitz, P. M., and P. J. Howarth. 2000. High spatial resolution remote sensing data for forest ecosystem classification: An examination of spatial scale, *Remote Sensing Environment*, 72(3), 268–290.

Treshow M., and M. R. Pack. 1970. Fluoride, In: Recognition of Air Pollution Injury to Vegetation, In: *A Pictorial Atlas,* Jacobson, J. S., and A. C. Hill (eds.). Air Pollution Control Assoc., Pittsburgh, PA, D1–D7.

Vogelmann, J. E., B. N. Rock, and D. M. Moss. 1993. Red edge spectral measurements from sugar maple leaves, *International Journal of Remote Sensing*, 14(8), 1563–1575.

Vogg G., R. Heim, J. Hansen, C. Schafer, and E. Beck. 1998. Frost hardening and photosynthetic performance of Scots pine (*Pinus sylvestris* L.) needles, I. Seasonal changes in the photosynthetic apparatus and its function, *Planta*, 204, 193–200.

Woodcock, C. E., and A. H. Strahler. 1987. The factor of scale in remote sensing, *Remote Sensing Environment*, 21, 311–322.

Yoder, B. J., and R. E. Pettigrew-Crosby. 1995. Predicting nitrogen and chlorophyll content and concentrations from reflectance spectra (400–2500 nm) at leaf and canopy scales, *Remote Sensing Environment*, 53, 199–211.

Zarco-Tejada, P. J., and J. R. Miller. 1999. Land cover mapping at BOREAS using red edge spectral parameters from CASI imagery, *Journal of Geophysical Research*, 104 (D22), 27,921–27,948.

Zarco-Tejada, P. J., J. R. Miller, G. H. Mohammed, T. L. Noland, and P. H. Sampson. 1999. Canopy optical indices from infinite reflectance and canopy reflectance models for forest condition monitoring: Applications to hyperspectral CASI data, In: Proceedings of the IEEE International Geoscience and Remote Sensing Symposium IGARSS, Hamburg, Germany, 1–4.

Zarco-Tejada, P. J, J. R. Miller, G. H. Mohammed, T. L. Noland, and P. H. Sampson. 2001. Scaling-up and model inversion methods with narrow-band optical indices for chlorophyll content estimation in closed forest canopies with hyperspectral data, *IEEE Transactions on Geoscience & Remote Sensing*, 39(7), 1491–1507.

10

Geology Applications

Using hyperspectral remote sensing to identify and map specific chemical and geometric patterns of the land is useful for studying geology, soil science, mining, land use, and hydrology. This information can be used to identify areas most likely to hold economically valuable deposits of minerals or petroleum.

Geologists can search vast territories looking for surface features that may indicate the presence of minerals in a certain area. Imaging spectroscopy allows specific absorption features, caused by chemical bonds in materials, to be mapped spatially. Materials maps are of minerals, mineral mixtures, vegetation (including species communities and vegetation communities maps), water, ice and snow, atmospheric gases, environmental materials, and man-made materials.

The two case studies in this chapter discuss the use of hyperspectral imagery to map and identify the mineralogy and chemistry of rocks and soils. In the first study, Hörig et al. (2001) used reference areas with known geometry and chemical properties, collected hyperspectral data using two different instruments, and compared the data to develop an optimum data correction and processing procedure.

Case Study 1: Detecting Hydrocarbons

In 1998, hyperspectral airborne scanners were used to directly detect hydrocarbon-bearing rocks and soils. This case study, called the Pro Smart Experiment, was organized by the German Aerospace Centre (DLR) and provided an opportunity to test the hyperspectral HyMap scanner designed by the Australian company Integrated Spectronics Ltd. Several European remote sensing organizations have used this experiment to evaluate the capability of this technology to clarify specific questions of geology, mining, land use, and hydrology.

In the Pro Smart Experiment, Hörig et al. (2001) used a large space in Berlin to prepare reference areas. The reference areas had a defined geometry and known chemical properties (e.g., sandy soil, oil-contaminated soil, grass, plastic tarpaulin). After collecting HyMap data for these areas, they compared their data with the spectra obtained simultaneously with the GER Mark V IRIS Infrared Intelligent Spectroradiometer. The HyMap data and the data provided by the Mark V IRIS were then used to develop an optimum data correction and processing procedure. The purpose of the processing was to unambiguously identify the hydrocarbon-bearing reference areas in the HyMap data.

The Pro Smart Experiment had two goals. The first goal of this case study was to provide a simple methodology for the routine detection and mapping of hydrocarbons. The second goal was to improve the general knowledge of the spectral properties of hydrocarbons.

Methods and Sensors

Test Field
The test field was prepared in the Spandau district of Berlin. A large parking lot formerly used by military vehicles provided ideal conditions for the reference areas. The concrete surface of the parking lot, the lawn, single trees, and a gravel-paved area were used as additional reference objects. The reference areas were used to define the spectral properties of the reference objects, to evaluate the effect of undersampling in the case of small targets, and to define the limits of the method in the case of slightly contaminated targets with low-intensity spectral features.

Field Spectroscopy
The spectral properties of the reference areas were determined with a GER Mark V IRIS Infrared Intelligent Spectroradiometer. The Mark V was operated with a 7×3 degree dual field of view in the spectral range from 0.385 to 2.548 μm (849 bands). The widths of its spectral band varied between 2 and 6 nm. The spectroradiometer was mounted on a cart so that it could be quickly moved between the reference areas during the HyMap flights. The spectroscopic data were recorded as ratio spectra (percent reflectance) and as radiance spectra [μW/(cm^2 nm sr)].

HyMap and HRSC-A Airborne Scanner

HyMap Scanner
Selected specifications of the HyMap scanner are shown in Table 10.1. The HyMap flights were carried out in September 1998. To obtain optimal spectral and spatial resolution, the HyMap scanner was flown at a relatively low altitude. The original objective to obtain HyMap datasets of 1 m GIFOV (ground instantaneous field of view) was not achieved because even the lowest operating speed of the aircraft was too high. To eliminate any gaps between the scan lines (i.e., no undersampling), the minimum GIFOV needed to be approximately 4 m with a flight altitude of 2,000 m. Therefore, two separate flights were made at altitudes of 2,200 m and 11,378 m.

The 2,200-m flight provided data with 4.4 m across-tract (5.5 m along-track) GIFOV and full coverage of the ground. The 1,137-m flight was a compromise. Spatial resolution was better at this altitude (2.27 m across-track GIFOV, 2.84 m along-track GIFOV), but because no further reduction of the aircraft's speed was possible and the integration time of the scanning system could not be changed, gaps occurred between the scan lines. This undersampling

Table 10.1 Selected Specifications of the HyMap Scanner

Spectral range	0.440–2.543 µm
Spectral bands	128
Spectral band widths	10–20 nm
IFOV	2.5 mrad (along track); 2.0 mrad (across track)
GIFOV	5 m at 2,500-m operating altitude
Signal-to-noise ratio	> 500:1

Source: Hörig B., F. Kühn, F. Oschütz, and F. Lehmann. 2001. *International Journal of Remote Sensing*, 22(8), 1413–1422. With permission.

resulted from the 2.84-m along-track GIFOV (i.e., width of each scan line) and the 7.67 m between the centers of the pixels of each scan line.

HRSC-A Camera

The HRSC-A camera is the airborne version of the High Resolution Stereo Camera (HRSC) originally designed for space applications. The HRSC-A has a charge-coupled device (CCD) system with 5,272 pixels per line, 7-µm pixel size (i.e., 10 cm at 2,500-m altitude), and 8-bit radiometric resolution (Scholten et al., 1999).

DLR's HRSC-A camera was operated simultaneously with the HyMap scanner. The HRSC-A camera provided high-resolution images of the area covered by HyMap. These images were used to verify the ground objects in the HyMap data and to rectify the HyMap images.

Results and Discussion

In most cases, visible to short-wave infrared spectra are recorded by a field or laboratory spectrometer as percentage reflectance of the incident radiation. Consequently, most spectra available from spectral databases are percentage reflectance spectra. The general advantage of percentage reflectance spectra is that they can be compared even if they were recorded at different times, at different geographic locations, and with different spectrometers.

In this case study, the HyMap pixel spectra were evaluated by comparison with the Mark V radiance spectra, which were made at the same time as the HyMap flights. This was done because the HyMap and Mark V instruments were calibrated in a similar way.

DLR applied radiometric corrections and correction for systematic errors to the radiance values of the HyMap data. Consequently, the digital number of each HyMap pixel was proportional to the absolute value of the radiance reflected by the ground and could be compared directly with the Mark V spectra of the respective reference area. Atmospheric corrections to the HyMap data were not necessary.

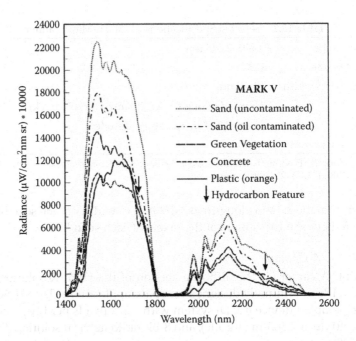

Figure 10.1 GER IRIS Mark V radiance spectra. (Hörig B., F. Kühn, F. Oschütz, and F. Lehmann. 2001. *International Journal of Remote Sensing*, 22(8), 1413–1422. With permission.)

Figure 10.1 shows Mark V radiance spectra of selected reference areas. As already demonstrated by Clutis (1989), the hydrocarbon-bearing reference objects are characterized by absorption maxima at wavelengths of 1,730 and 2,310 nm. In this case study, plastic objects (e.g., plastic tarpaulin) showed a relatively sharp absorption maximum at 1,730 nm, whereas this peak was less prominent in the spectra of oil-bearing soils and rocks. The question was whether this less prominent absorption peak (or radiance minimum) could be recognized in the HyMap pixel spectra, despite noise produced by the atmosphere between the scanner and the ground.

Figure 10.2 shows the HyMap pixel spectra of the same reference areas as in Figure 10.1. The spectra were calculated using ENVI image processing. The same maxima/minima that are characteristic of hydrocarbons are present in both plots. Although less prominent, they are significant enough for hydrocarbon-bearing materials to be detected when the pixel spectra are evaluated. However, efficient mapping of the locations of hydrocarbons requires image processing capable of accentuating all pixels with such absorption maxima. All pixel spectra and images were from data obtained at 1,137-m flight altitude.

The best results were obtained for the 1,730-nm maximum/minimum defined by HyMap short wave infrared (SWIR)-1 bands 21 (1,668.22 nm), 26 (1,729.31 nm), and 31 (17,88.98 nm). These bands were used for false color

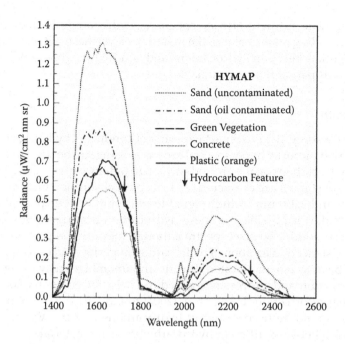

Figure 10.2   HyMap spectra. (Hörig B., F. Kühn, F. Oschütz, and F. Lehmann. 2001. *International Journal of Remote Sensing*, 22(8), 1413–1422. With permission.)

composite (FCC) images based on SWIR-1 bands 21/red, 26/green, and 31/blue combined with linear contrast stretching. The HyMap SWIR-1 bands 21, 26, and 31 are located in a narrow portion (100 nm wide) of the electromagnetic spectrum. Consequently, most of the surface materials appear gray because they show similar spectral characteristics in this portion of the spectrum. Only the hydrocarbons appear colored due to their significant absorption features within the same narrow portion.

This procedure combined with linear stretching led to grayscale images with colored depiction of all hydrocarbon-bearing materials. Consequently, the oil-contaminated soil, the plastic tarpaulin, and the plastic sheet appeared pinkish. The intensity of the color of the large oil-contaminated reference areas was proportional to the oil content.

The percentage of the smaller areas covered by pixels could not be determined. However, the 2.27-m GIFOVs covering or touching the small oil-contaminated reference areas (2 m × 2 m and 1 m × 1 m) appear slightly pinkish. Additionally, the lower limit on the aircraft speed caused undersampling.

Oil-contaminated soil and other materials made of hydrocarbons could be distinguished in color-composite images based on HyMap bands in the visible and the near-infrared portions of the spectra. In these parts of the spectrum, the colors of the objects help distinguish, for example, between plastic, artificial grass, roofing felt, and oil-contaminated soil. The Vis/NIR (visible to near infrared) HyMap spectra of hydrocarbon-bearing materials differ depending on the color of the material. Nevertheless, it was impossible to differentiate between hydrocarbons and non-hydrocarbons on the basis of the Vis/NIR part of the spectrum.

Conclusion

In this case study, the Pro Smart Experiment conducted by Hörig et al. (2001) showed that airborne hyperspectral remote sensing can be used to efficiently detect hydrocarbons on the ground surface. In radiance spectra, hydrocarbon-bearing substances are characterized by typical absorption maxima at about 1,730 nm and 2,310 nm. A high signal-to-noise-ratio hyperspectral imaging system makes it possible for these hydrocarbon-bearing substances to be recognized within the pixel spectra although they are less pronounced.

The existence and location of hydrocarbon materials and oil-contaminated soil can be detected and located directly and unambiguously using a simple data correction and processing procedure; it is not necessary to apply an atmospheric correction. Hörig et al. (2001) found that the SWIR part of the spectrum could be used to detect hydrocarbons and the Vis/NIR part to distinguish between different hydrocarbon-bearing materials.

A prerequisite for efficient mapping of hydrocarbons was the availability of a hyperspectral sensor with a high signal-to-noise ratio (HyMap) and simultaneous field spectroscopy data for well-defined reference areas so that the spectral signatures of the target objects could be defined. Using this approach, it should be possible to detect and map any objects (i.e., rocks, minerals, soils) directly in the same simple way, even if they are character-ized by low-intensity radiance spectrum features.

Case Study 2: Mapping Alteration Zones Associated with Gold Mineralization

In the second case study, Ferrier and Wadge (1996) used data from the SWIR part of the spectrum to map the general location and amount of hydrous alteration minerals at a site in southern Spain. Being able to locate these minerals, which are associated with gold mineralization, is important for exploration geology.

Introduction

When fluid flow processes substantially alter the mineralogy and chemistry of host rocks, economically valuable mineralization can occur. This altera-tion can produce distinctive assemblages of minerals that vary according to the location, degree, and longevity of those flow processes. But it doesn't help anyone's economy until the location of these minerals can be determined.

Once the minerals are exposed at the surface of the Earth, this alteration can sometimes be mapped as a zonal pattern around a core of the highest-grade alteration and greatest economic interest.

Narrow-band sensors such as the AVIRIS imaging spectrometer are capable of detecting spectral absorption in the visible to short-wave infrared specific to some individual minerals (Vane and Goetz, 1988). If these minerals are indicative of the type of alteration and are present in sufficient quantities at the surface from which solar radiation can be reflected to the sensor, then imaging spectrometers offer the prospect of a valuable, additional source of data for the exploration geologist.

This section discusses an area of gold mineralization at Rodalquilar in southern Spain. Rodalquilar is the site of the Transaccion gold mine, which was worked intermittently during the 20th century. Traditional maps of alteration "zones" were constructed for Rodalquilar based on field survey and laboratory work (Arribas et al., 1989; Hernandez et al., 1989). For this case study, Ferrier and Wadge (1996) then applied unmixing analysis, spectral curve matching, and absorption feature identification/analytical techniques to the data and compared the ground-based alteration maps with the results of the AVIRIS mapping.

Geological Background

Rodalquilar lies in the Sierra de Gata structural block of southern Spain, which is the most southeasterly of the internal zones of the Betic orogen. The Sierra de Gata trends northeasterly, is bounded to the northwest by a complex left lateral strike slip fault zone, and is characterized by calcalkaline volcanism of Miocene age, whose deposits dominate the geology at the southern end. Between 14 and 11 Ma silicic magmatism developed beneath earlier volcanic rocks comprising hornblend- and pyroxene-bearing andesites.

Two Valles-type calderas, the Las Frailes Caldera (Cunningham et al., 1989) and the Rodalquilar Caldera further north (Rytuba et al, 1989), were identified. The deposits of these calderas include rhyolitic to dacitic ignimbrites and resurgent domes. Post-caldera volcanism reverted to the eruption of andesite lavas.

The epithermal gold mineralization at Rodalquilar is of the acid sulphate type (Heald et al., 1987), hosted by rhyolitic ignimbrite deposits and domes within the caldera. There are economic ores of gold-alunite and lead-zinc-silver-gold veins, principally concentrated in ring and radial fractures around the western margin of the Cinto caldera (Transaccion mine), and alunite deposits in the Los Tollos area at the northeastern margin of the Rodalquilar caldera (Arribas et al., 1989)

The gold-bearing ores were concentrated in a few square kilometers at the center of the zone of alteration that filled much of the Rodalquilar Caldera and extended to a depth of 200 m below the paleosurface. Alteration was proved by drilling depths of over 900 m. Similar low- to high-grade alteration zones were encountered upward and inward toward the Cinto areas, which is interpreted as the focus of upwelling hot acidic fluid.

Spectral Character of Alteration Assemblages

Reflectance spectroscopy of rocks and minerals can give useful diagnostic information on their elemental and mineralogical composition (Hunt, 1977). The remote application of this technique (e.g., Vane and Goetz, 1988) from airborne sensors finds its most obvious application in the mapping of hydrothermally altered rocks (e.g., Krohn, 1986; Rast, 1991). This is because many of the alteration minerals that were produced have distinctive absorption features caused by the presence of OH and other hydroxyl bonds, Mg-OH, and Al-OH, particularly in the SWIR part of the spectrum (2,000–2,400 nm).

The main diagnostic minerals of the alteration zones at Rodalquilar were quartz-alunite, pyrophyllite, kaolinite, illite, and vermiculite. Laboratory reflectance spectra, with wavelength positions, depths, and number of absorption features mainly in the 1,400-, 1,700-, 1,900-, and 2,200-nm regions of the spectra, should allow researchers to determine the minerals present and then form the basis for mapping the zonation from an equivalent remote sensor such as AVIRIS.

However, a number of factors inhibit mineral-identification mapping in this case. The AVIRIS data themselves are less accurate than laboratory spectra. The spectral resolution is coarser (10 nm as opposed to 1 to 2 nm or less for laboratory spectrometers), is noisier, and has an effective loss of signal in the 1,400-nm and 1,900-nm regions where atmospheric water absorbs the radiation.

The nature of the ground surface further affects the value of the reflected signal. The overall brightness or albedo of the signal (and hence the magnitude of any absorption features) depends on the local relief and exposure of altered rock. Therefore, at Rodalquilar, the "active" mine surface workings tend to dominate the areas from which good spectral information can be extracted, and the workings may not represent in situ rock (e.g., dumps).

An iron-rich thin soil is common at Rodalquilar, and a soil sheet wash, down steep slopes, tends to mask the rock signal. Patchy senescent vegetation practically obscures some areas. The proportion of these hydrous minerals in the rocks is small—the original rhyodacites had over 70% silica and alteration has introduced even more—and some zones contain intimate mixtures of more than one "diagnostic" mineral. Finally, the 20-m by 20-m field of view of the AVIRIS sensor means that, in addition to the intimate hand-specimen level of mineral mixing, the signal received at the plane is an integrated mix of contributions from different rock types at the scale of 10 m and also from many of the other unwanted effects listed above.

AVIRIS Data

The AVIRIS consists of four separate spectrometers. Signal-to-noise ratios for the first three spectrometers were good, averaging about 100 for bright targets and in the range of 100–50 for dark targets. Unfortunately, the D-spectrometer that recorded the SWIR produced very noisy data throughout the MAC EUROPE '91 campaign, apparently because of a damaged optical fiber cable. The signal-to-noise ratio for the D-spectrometer, therefore, was poor, averaging about 12 in the central part of the range (near 2,200 nm) but falling off to 5 or less at either limit for bright targets and staying below 5 for dark targets throughout the range. Therefore, the part of the spectrum most useful for identifying hydrous mineral contained no data of value for dark targets in our data although there was some useful information from areas of high albedo.

Retrieval of Apparent Surface Reflectance

Empirical Line Method

The empirical line method for the recovery of surface reflectance has been described in a number of papers (e.g., Roberts et al., 1986; Conel et al., 1987). The empirical line method is based on the following simplified equation:

$$DNb = \rho(\lambda)Ab + Bb$$

where DNb equals the digital number for a given pixel in band b, $\rho(\lambda)$ equals the reflectance of the surface materials within that pixel at the wavelength λ of band b, Ab equals the multiplicative term that affects the DN (transmittance and instrumental factors) and Bb equals the additive term (primarily atmospheric path radiance and instrumental offset, i.e., dark current).

Four homogeneous ground targets of reasonable size and varying brightness were chosen, and the appropriate AVIRIS pixels were identified from the image data. The single field of view IRIS spectrometer (SIRIS) ground spectra for each target were then averaged and subsequently convolved to the bandwidths of the AVIRIS instrument. The AVIRIS DN were then plotted against the SIRIS-measured surface reflectance values for each target. The best fitting plot was determined using a least-squares fitting technique. The slope of the resulting line is the gain for that band and the y-intercept is the offset. Better statistics can be obtained by using more surface calibration targets.

The empirical line method was seen to be critically dependent on the quality, number, and localities of the ground spectra. This method is also very sensitive to topographic variations with a large error introduced in areas of even moderate relief (Conel et al., 1987).

The modeled spectra using radiosonde data gave the best overall results. However, the essential requirement for accurate topographic information to correct the total water path could be an expensive and limiting factor in many areas.

The spectra modeled by radiative transfer methods are similar and correlate highly with the library spectra. In this case study, Ferrier and Wadge (1996) found that water path profiles retrieved from the imaging spectrometer data satisfactorily corrected the remotely sensed data for atmospheric effects. This obviates the requirement for radiosonde data to be acquired at the time of the overflight and subsequently corrected for topographic variation. It also eliminates the requirement for ground spectra to be obtained at a number of targets at the time of the overflight.

Results

Spectral Matching

As stated earlier, areas of high albedo have SWIR signal-to-noise ratios of about 12—around 2200 nm. This is sufficient to match AVIRIS SWIR spectra to library spectra at modest match scores of 30–50 (0–200 full range) and delineate areas of hydrous minerals obviously associated with the acid-sulphate-type mineralization. Kaolinite/dickite (the strongest), pyrophyllite, calcite, gypsum, muscovite, and montmorillonite matches were found. However, the lack of identification of alunite, which does have absorptions in the SWIR and which can be identified from the C-spectrometer data, throws considerable doubt on the value of individual pixel identifications by Automatic Intelligent Material Location and Identification System (AIMLIS). In this case study, most of the matches were only indicative of the presence of an absorption feature around 2,200 nm due to either Al-OH or Mg-OH although some of the highest kaolinite/dickite identifications at match scores of 70 could have been genuine (Wadge et al., 1993).

Spectral Unmixing

The Rodalquilar scene was unmixed using about 130 of the bands from spectrometers A–C condensed to 13 principal component bands and 12 end-members selected from the image, including four "alteration" endmembers. Output of the mixing was restricted to produce only the two dominant components in each pixel, and the results were presented as maps in which individual alteration endmembers were the dominant proportion. Three of the four endmembers were suggested to be alunite, illite, and kaolinite. The endmembers clustered around the heart of the alteration zone, but apart from the correlation with alunite, they showed no simple correlation to the alteration zones mapped on the ground. Further attempts at a more detailed unmixing analysis on a subscene that was centered on the mine were made using various band combinations and endmembers, but these did not prove successful, probably because the endmember spectra were too similar.

Conclusion

A comparative assessment of various methods of calculating apparent reflectance values for AVIRIS data shows that radiative transfer modeling using image-derived water vapor abundances performs almost as well as methods requiring radiosonde data.

Unlike earlier researchers who have also used data from the 2,000–2,400-nm range to detect alunite, in this case study, Ferrier and Wadge (1996) demonstrated the ability of AVIRIS data to map the spatial distribution of alunite at Rodalquilar using only the combined depths of absorption features at 1480 and 1,760 nm. In addition, SWIR data (2,000–2,400 nm) at signal-to-noise ratios of 12–5 were sufficient to map the general location and amount of hydrous alteration minerals at Rodalquilar but not to identify individual zone minerals. Finally, detailed ground and laboratory analysis of the nature of mineral assemblages and their spectral integration are needed to understand high-quality imaging spectrometry data of alteration zones.

References

Arribas, A., Jr., J. J. Rytuba, R. O. Rye, C. G. Cunningham, M. H. Podwysocki, W. C. Kelly, A. Arribas, Sr., E. H. McKee, and J. G. Smith. 1989. Preliminary study of the ore deposits and hydrothermal alteration in the Rodalquilar Caldera Complex, southeastern Spain. U.S. Geological Survey Open-File Report 89-327, Reston, Virginia, 39 pp.

Clutis, E. A. 1989. Spectral reflectance properties of hydrocarbons: remote sensing implications, *Science*, 245, 165–168.

Conel, J. E., R. O. Green, G. Vane, C. J. Bruegge, and R. E. Alley. 1987. AIS-2 radiometry and a comparison of methods for the recovery of ground reflectance. Proceedings of the 3rd Airborne Visible/Infrared Imaging Spectrometer (AVIRIS) Workshop, Pasadena, CA, 20–21 May 1987, edited by R. O. Green, JPL Publication 87-30 (Pasadena, CA: Jet Propulsion Laboratory), 18–47.

Cunningham, C. G., A. Arribas, Jr., J. R. Rytuda, and A. Arribas, Sr. 1989. Evolution of the Los Frailes caldera, Cabo de Gata volcanic field. U.S. Geological Survey Open-File Report 89-325, Reston, VA, 20 pp.

Ferrier G., and G. Wadge. 1996. The application of imaging spectrometry data to mapping alteration zones associated with gold mineralization in southern Spain, *International Journal of Rock Mechanics and Mining Sciences and Geomechanics Abstracts*, 33(7), 308A–308A.

Heald, P., N. K. Foley, and D. O. Hayba. 1987. Comparative anatomy of volcanic-hosted epithermal deposits: Acid-sulphate and adularia-sericite types, *Economic Geology*, 82, 1–26.

Hernandez, P. A., P.A. Garcis-Estrada, and P. N. Cowley. 1989. Geologic setting, alteration, and lithogeochemistry of the Transaccion epithermal gold deposits, Rodalquilar mining district, southeastern Spain, *Transactions Institute of Mining and Metallurgy*, 98, B78–B82.

Hörig B., F. Kühn, F. Oschütz, and F. Lehmann. 2001. HyMap Hyperspectral Remote Sensing to Detect Hydrocarbons, *International Journal of Remote Sensing*, 22(8), 1413–1422.

Hunt, G. R. 1977. Spectral signatures of particulate minerals in the visible and near-infrared, *Geophysics*, 42, 501–503.

Krohn, M. D. 1986. Spectral properties (0.4–2.5 microns) of selected rocks associated with disseminated gold and silver deposits in Nevada and Idaho, *Journal of Geophysical Research*, 91, 767–783.

Rast, M. 1991. Imaging spectroscopy and its application in spaceborne systems. ESA SP-1144 (Paris: European Space Agency), 144 pp.

Roberts, D. A., Y. Yamagushi, and R. J. P. Lyon. 1986. Comparison of various techniques for calibration of AIS data. Proceedings of the 2nd Airborne Visible/Infrared Imaging Spectrometer (AVIRIS) Workshop, Pasadena, CA, 4–5 June 1990, edited by G. Vane and A. F. H. Goetz, JPL Publication 86-35 (Pasadena, CA: Jet Propulsion Laboratory, 21–30.

Rytuba, J. J., A. Arribas, Jr., C. G. Cunningham, E. H. McKee, J. G. Smith, and A. Arribas, Sr. 1989. Evolution of the Rodalquilar caldera complex and associated gold-alunite deposits, Cabo de Gata volcanic field, Southeastern Spain. U.S. Geological Survey Open-File Report, 89-326, Reston, VA, 18 pp.

Scholten, S., S. Sujew, F. Wewel, J. Flohrer, R. Jaumann, F. Lehmann, P. Pischel, and G. Neukum. 1999. The high resolution stereo camera HRSC-A—digital 3-D image acquisition; photogrammetric processing and data evaluation, In: Proceedings, Joint Workshop Sensors and Mapping from Space 1999, Institut fur Photogrammetrie und Ingenieurvermessung, Universitat Hannover, Hannover, No. 18.

Vane, G., and A. F. H. Goetz. 1988. Terrestrial imaging spectroscopy, *Remote Sensing of Environment*, 24, 1–29.

Wadge, G., G. Ferrier, S. Mackin, J. McM. Moore, S. Manthripragada, J. G. Liu, and R. Murphy. 1993. Mapping hydrous pathfinder minerals associated with gold mineralization in southern Spain. Proceedings of the 25th International Symposium, Remote Sensing and Global Environmental Change, Graz, Austria, 4–89 April 1993 (Ann Arbor: ERIM), 1220–1230.

Index

Printed in the United States
by Baker & Taylor Publisher Services